Electrical Inspection, Testing and Certification

Updated in line with the 18th edition of the wiring regulations.

This book is an essential guide to the City & Guilds 2391-50 and 51: Initial Verification and Certification of Electrical Installation and Periodic Inspection and Testing, also C&G 2391-52: an amalgamation of Initial Verification and Periodic Inspection and Testing of electrical installations.

There is a full coverage of technical and legal terminology used in the theory exams; including the structure of exam questions and their interpretation. By running through examples of realistic exam questions in a step-by-step fashion, this book explains how to decode the questions to achieve the most suitable response from the multiple-choice answers given.

This book is ideal for all electricians, regardless of their experience, who need a testing qualification in order to take the next step in their career.

Michael Drury has worked in the electrical industry for over 50 years, and is currently working as a freelance electrical installation lecturer and as an on-site constructor, designer and inspector. Michael has worked in FE for 25 years and has also been employed as a contractor in the United Arab Emirates and Saudi Arabia.

Electrical Inspection, Testing and Certification

A Guide to Passing the City and Guilds 2391 Exams

Third Edition

Michael Drury

Routledge
Taylor & Francis Group

LONDON AND NEW YORK

Third edition published 2020
by Routledge
2 Park Square, Milton Park, Abingdon, Oxon, OX14 4RN

and by Routledge
52 Vanderbilt Avenue, New York, NY 10017

Routledge is an imprint of the Taylor & Francis Group, an informa business

© 2020 Michael Drury

First edition published by Routledge 2016
Second edition published by Routledge 2018

British Library Cataloguing-in-Publication Data
A catalogue record for this book is available from the British Library

Library of Congress Cataloging-in-Publication Data
A catalog record has been requested for this book

ISBN: 978-0-367-43027-6 (hbk)
ISBN: 978-0-367-43026-9 (pbk)
ISBN: 978-1-003-00077-8 (ebk)

Typeset in Sabon
by Taylor & Francis Books

Table of contents

Contents

Introduction

The process of inspecting and testing electrical installations has evolved over a number of decades, as electrical equipment has become more prolific and sophisticated. Consequently, the demands on the practising electricians are ever increasing both at the skills level and those imposed by accumulative regulative liabilities.

In order to cope with these increasing demands there is a constant urgency to develop and maintain appropriate and necessary skills to meet the pressures imposed on them by *Statutory and Non-Statutory Regulations*; which have sensibly been introduced and frequently up-dated, in order to reduce or possibly eliminate the number of fatalities, serious injuries in the work place and the domestic environment, also to prevent damage to property.

Safety has always been, and must be, at the forefront of all practising electricians' work ethos, which oddly can be a problem when they sit the *City & Guilds 2391 theory exams*; simply because they have a natural reaction to solve or rectify an electrical fault, which may have been exposed during an inspecting and testing routine. Consequently, if a candidate (inspector/electrician) is presented with a question on a possible fault, defect or omission, this natural reaction may kick in. Unfortunately this type of response is *not* expected from the candidate during the theory exam.

The answer given will ultimately depend on the mode of the inspection paper, whether it's *Initial Verification, Periodic Inspection*, or a combination of both, the response will be entirely different.

Accordingly candidates must always bear in mind their position as an *inspector* when sitting the *theory exam(s)*; where they are expected to carry out the inspection and testing of an installation regardless of the type of inspection and *give an account on its condition*. Subsequently, it is paramount for the candidate (inspector) to understand the testing and inspection procedures, with all the associated certification and schedules, as laid down in the current version of *BS 7671 Wiring Regulations* and the *IET Guidance Note 3 Inspection & Testing*.

The inspector must also be fully aware of the safety procedures coupled with inspection and testing, moreover the action to be taken if, in the inspector's professional opinion, an installation is unsafe or does not conform to the designer's criteria or BS 7671.

Consequently the objective of this book is to assist and review the working practices of the skilled electricians to ensure they are familiar with the details of both the statutory and non-statutory regulations, to assist in the decoding of questions and scenarios posed by *City & Guilds*, thereby gaining a greater understanding of the inspecting and testing of an electrical installation with the ultimate objective of ensuring an electrical installation is safe to use.

In order to achieve this objective the installation must comply with the designer's criteria; its construction should conform to all aspects of BS 7671 and it has been inspected and tested to ensure it is in a satisfactory condition where it can be used safely.

A SUMMARY OF THE CITY & GUILDS 2391

The City & Guilds 2391 consists of three stand-alone Units, they are:

- C&G 2391-50 *Initial Verification of Electrical Installations.*
- C&G 2391-51 *Periodic Inspection and Testing of Electrical Installations.*
- C&G 2391-52 *Initial and Periodic Inspection and Testing.*

There are numerous similarities between each Unit; however candidates will be expected, depending on the Unit taken, to have a detailed understanding of the following:

- The *requirements for either an Initial Verification or a Periodic Inspection* of electrical installations.
- The *safety management procedures for either an Initial Verification or a Periodic Inspection* when undertaking inspections and testing of electrical installations.
- The *requirements for carrying out either an Initial Verification or a Periodic Inspection* of an electrical installation.
- The *requirements for testing either an Initial Verification or a Periodic Inspection* of an electrical installation.

1
Statutory and non-statutory regulations

INTRODUCTION

The legal aspects of Statutory and Non-Statutory Regulations are probably the least consulted and possibly the most neglected components within the Inspection and Testing domain; whereas they should be regarded as the bedrock of the Inspection and Testing process.

When an inspector, generally the installation electrician on relatively small installations or as a signatory for larger installations, issues an *Electrical Installation Certificate (EIC)*, it is a legally binding document, whereas an *Electrical Installation Condition Report (EICR)* is, as the title of the document implies, a *report* on the condition of an installation. Nevertheless both documents will depend upon the professionalism, knowledge, experience and skills of the inspector.

1.1 Legal responsibilities
The relationship between statutory and non-statutory regulations

i) Statutory documents
    a) *Health & Safety at Work etc Act 1974*
    b) *Electricity at Work Regulations 1989.*
ii) Non-Statutory documents
    a) *BS 7671 Requirements for Electrical Installation Wiring Regulations*
    b) *IET On-Site-Guide*
    c) *IET Guidance Note 3 Inspection & Testing*
    d) *HSE Guidance GS 38 Electrical Test Equipment for use on Low Voltage Electrical Systems (4th Edition).*

The *Electricity at Work (EAW) Regulations 1989* is an "enabled" document, which means it has not taken the long tedious passage through the parliamentary procedure, but has been placed on the Statute Books under the umbrella of the

Health & Safety at Work etc Act 1974, which confirms its status as a statutory document, therefore it can be used in a court of law; whereas *BS 7671 Wiring Regulations* is a *code of practice*; nevertheless they may also be referred to in any judiciary process (see *BS 7671 Regulation 114 page 16*).

The legal consequences which can result from an inspector deviating from, or failing to comply with, *BS 7671 Wiring Regulations* are immense, more so if the inspector's dereliction results in a fatality or a serious injury; then the inspector may be subjected to the full weight of the legal system, strongly pursued and enforced by the *Health & Safety Executive (HSE)*. Therefore it is crucial for the inspector to actually *read* the *Regulations*, not just simply skim read or just take a brief perusal; these *Wiring Regulations* are essential reading. Moreover they are indispensable to an inspector because they are there to ensure, where reasonably practical, an electrical installation, whether it is domestic, commercial or industrial, is electrically safe. Failure to adhere to *BS 7671 Wiring Regulations* may result in legal penalties (see IET *Wiring Matters* Summer Issue 2012 and ESC *SwitchedON* Issue 24 Spring 2012, re: *Unlawful Killing Verdict*).

1.2 The *memorandum of guidance on the electricity at work regulations 1989*

This *Memorandum* is obligatory for all work-related electrical personnel for their own safety and to ensure precautions are taken to safeguard the safety of others from the risk of injury or death.

The *Memorandum* is primarily a guide to assist *duty holders* meet and fulfil the requirements of the *Electricity at Work Regulations 1989* in the form of "technical and legal guidance". The document defines and explains the *duties* imposed on a *duty holder* in *Regulation 3 paragraph 55* as one who must *"comply with the provisions of these Regulations in so far as they relate to matters which are within his control"*.

Therefore, because the inspector's title and status are defined, within a statutory document, which is directly linked to the judiciary, a *City & Guilds* question could simply ask the candidate for the inspector's title and/or status in LAW, which are duty holder and competent person respectively (see *Regulations 3 & 16*).

There is always an element of risk when carrying out electrical testing; consequently an inspector has a duty of care to safeguard his/her own safety as well as that of others. In an endeavour to reduce or eliminate possible hazards, when inspecting and testing, there are mandatory Regulations which the inspector must observe. Possibly the most significant are:

- **Regulation 11** – Means for protecting from excess of current: *Efficient means, suitably located, shall be provided for protecting from excess of current every part of a system as may be necessary to prevent danger.*

In other words, the inspector should ensure a circuit's protective device is suitable for protection against overloads and short circuits, where applicable, earth faults.

- **Regulation 12** – Means for cutting off the supply and for isolation: *where necessary to prevent danger, suitable means including, where appropriate, methods of identifying circuits shall be available for a) cutting off the supply of electrical energy to any electrical equipment; and b) the isolation of any electrical equipment.*
- **Regulation 13** – Precautions for work on equipment made dead: *Adequate precautions shall be taken to prevent electrical equipment, which has been made dead in order to prevent danger while work is carried out on or near that equipment, from becoming electrically charged during that work if danger may thereby arise* (use of safety locks for locking-off appropriate circuits and isolators).

Regulations 12 and 13 are directly associated with the safe isolation.

Failure to carry out safe isolation, in the event of a fatality or serious injury, can lead to criminal proceedings being taken out against the inspector. If such an incident should occur the inspector must prove he or she had taken *all reasonable steps* and *exercised all due diligence* to avoid committing an offence under regulations 11, 12 and 13.

- **Regulation 29** – Defence: *In any proceedings for an offence consisting of a contravention of regulations 4(4), 5, 8, 9, 10, 11, 12, 13, 14, 15, 16 or 25, it shall be a defence for any person to prove that he took all reasonable steps and exercised all due diligence to avoid the commission of that offence.*

Regulation 29 applies *only in criminal proceedings*. It provides a defence for a dutyholder who can establish that they took *all reasonable steps* and *exercised all due diligence* to avoid committing an offence under regulations 4(4), 5, 8, 9, 10, 11, 12, 13, 14, 15 or 16.

Related topic questions

The types of questions which could be asked on this topic are generally straightforward, for example:

Q1. Which Statutory document identifies the title and status of an inspector?
  a) Health and Safety at Work etc Act 1974
  b) Electricity at Work Regulations 1989
  c) Electricity at Work Act 1989
  d) Electricity Safety, Quality and Continuity Regulations 2002.

Q2. Define the inspector's title in law.
  a) Duty Holder
  b) Skilled Person
  c) Competent Person
  d) Inspector.

Q3. Define the inspector's status in law.
  a) A Duty Holder
  b) A Skilled Person
  c) A Competent Person
  d) An Inspector.

1.3 *BS 7671 wiring regulations*

BS 7671 Wiring Regulations is a *code of practice* which *applies to the design, erection and verification of electrical installations.* It is a set of regulations and guidelines issued by the *Institution of Engineering & Technology (IET)*, a professional society with over 150,000 members.

The regulations and guidelines have been developed over numerous decades to the present-day standards, and are frequently updated to meet the ever evolving technical developments. Subsequently the objective of BS 7671 is to ensure, where reasonably practical, that an electrical installation is intrinsically safe. Consequently it is the responsibility of the inspector to confirm whether the *Regulations* have been adhered to and the installation *is* electrically safe. Therefore it is critical for the inspector to be fully conversant with *BS 7671 Wiring Regulations*, not just Part 6 of this document, simply because it is headlined as *Inspection & Testing*!

1.4 Inspector's status: differentiating between the terms *in law* and *legal*

Differentiating between the concepts of "in law" and "legal" can prove to be problematic simply because of their similarity. Grasping the fundamental difference can be critical because it is the inspector's *duty of care* to ensure an installation is electrically safe, where reasonably practical. To achieve this objective the inspector needs to consult the *Non-Statutory* set of *Regulations*, which can be supported by *Statutory Regulations* and enforced in a Court of Law.

Therefore what is the difference between "in law" and "legal"?

As previously explained, the term "law" is directly linked to a statutory document, namely the *EAW Regulations*, which can be used and enforced in a court of law, whereas "legal" can be defined as "authorised" or "approved".

BS 7671 is an approved code of practice under Section 16 of the *Health and Safety at Work etc Act 1974(HSWA)*, and its *legal* status is fully explained in Section 17 of the 1974 Act and therefore authorised. Furthermore, BS 7671 states:

> *Electrical Installation Certificates, Electrical Condition Reports and Minor Electrical Installation Certificates shall be "complied and signed or otherwise authenticated by a skilled person,* <u>competent</u> to verify that the requirements of the Standard have been met".

Note: The *Standard* is BS 7671.

Therefore the *legal title and status* of an inspector is a *skilled person (electrically) who is competent in such work*, that is, inspecting and testing of electrical installations.

Note: The term "work" in the statement: *competent in such work* can be interpreted as the type of electrical discipline/speciality an electrician is undertaking or involved with, for example domestic installations or electrical maintenance.

Accordingly, *City & Guilds* could generate a variety of questions, in various forms, to ensure the inspector is fully conversant with his/her title and legal status, responsibilities, knowledge and understanding of *BS 7671 Wiring Regulations*, which is the document that confirms the title and status of the inspector as: *a skilled person (electrically) competent in such work* (see BS 7671 Part 6 Reg. 644.5 page 236 & Reg. 653.5 page 238).

Related topic questions

Q4. The legal title of an inspector is:
a) A skilled person
b) A skilled person (electrically)
c) A competent person
d) An electrician.

Q5. The legal status of the inspector when carrying out Inspection and Testing is:
a) A competent person
b) A skilled person
c) A skilled person (electrically) competent in such work
d) An electrician.

Q6. State the legal status of the signatories of the Electrical Installation Certificate (EIC).
a) A competent person
b) Skilled persons competent to verify that the requirements of the Standard have been met
c) A skilled person(s) competent in such work
d) A designer.

1.5 *IET guidance note 3 inspection & testing*

The IET's *Guidance Note 3: Inspection & Testing is essential reading for all City & Guilds (C&G) 2391* candidates; although the booklet does not ensure compliance with BS 7671 it does however explain the *requirements* of BS 7671, which must always be consulted to confirm compliance with the Standard.

1.5.1 Responsibility of the inspector

It is the *responsibility of the inspector*, when conducting either *Initial Verification* and/or *Periodic Inspection*, to:

- ensure no danger occurs to any person or livestock and property is not damaged;
- compare the Inspection and Testing results with the design criteria, where available, BS 7671 and/or previous records, as appropriate;
- confirm compliance or non-compliance with BS 7671;
- make a professional/engineering view of an installation and report on its condition.

IET Guidance Note 3 is divided into two broad Sections. They are:

i)   Initial Verification
ii) Periodic Inspection and Testing.

i) Initial Verification

Initially, the title of this Section gives an illusion that it only applies to C&G 2391-50 or 52 candidates. Unfortunately it is slightly misleading. There are a number of common topics which link *all* C&G 2391 candidates and they should be aware of this factor.

For example: the types of tests, the instruments used for each test and the relevant testing procedure are all common to C&G 2391 candidates; although the testing sequence for a Periodic Inspection may *not necessarily* be the same as for an Initial Verification.

The main sub-sections within the Initial Verification Section are:

- Certificates
- Purpose of Initial Verification
- Required Information
- Frequency of Subsequent Inspections
- Initial Inspection
- Initial Testing.

1.5.2 Purpose of initial verification

This is to confirm a new installation, in addition to the existing installation or alteration to the existing installation, complies with BS 7671 in terms of *design and construction*, which is achieved through Inspection and Testing during and on completion of the installation (1st and 2nd fix inspecting and testing process).

Before any testing takes place, the installation's electrical equipment must be inspected for the following:

- it complies with British and harmonised (equivalent) Standards;
- it has been correctly selected and erected;
- there are no signs of visual damage or defects (also see BS 7671 Regulation 642.2 page 230).

1.5.3 Required information

The following information should be made available to the inspector, and recorded on both the *Electrical Installation Certificate (EIC)* and the *Electrical*

Installation Condition Report (EICR) with the exception of Maximum Demand, which is only recorded on the EIC type of earthing (TN-S, TN-C-S or TT).

- Number and type of live conductors
- Nature of Supply Parameters
- Supply Protective Device.

(see BS 7671 *Part 3 Chapters 311 & 312 pages 43-47, Appendix 6 pages 463 & 474 and GN3 page 84*).

Nature of Supply Parameter elements are:

- Nominal voltage and frequency
- Prospective fault current
- External earth fault loop impedance.

Related topic questions

Q7. State the purpose of Initial Verification.
a) The installation complies with BS 7671
b) The installation complies with the designer's criteria
c) The installation complies with the design and construction aspects of BS 7671
d) The installation is safe to use.

Q8. There are five supply characteristics which are recorded on an Electrical Installation Certificate (EIC). State the value of the nominal voltage which must be documented.
a) 230 volts
b) The measured value taken on the incoming side of the installation's supply
c) The value given by an enquiry to the DNO
d) The value given on the installation's schedule.

(ii) Periodic Inspection and Testing

This Section is applicable to C&G 2391-51 and 52. The main sub-headings are:

1 Its Purpose
2 Why is it Necessary?
3 Required Information
4 Frequency of Periodic Inspection
5 Setting Inspection and Testing Samples
6 Periodic Testing.

1 The *Purpose of Periodic Inspection and Testing* is to determine, so far as is reasonably practical, whether an installation is in a satisfactory condition to continue to be used safely.

2 *Why is Periodic Inspection and Testing necessary?* It is necessary because electrical installations deteriorate over time due to, for example, overloading, wear and tear. There are other factors which can determine the necessity for a Periodic Inspection and Testing. For example:
   - a legal requirement to ensure an installation is electrically safe;
   - other interested bodies such as licensing authorities, insurance companies and mortgage lenders. These organisations need to know if the property, irrespective of the type of installation, is electrically safe thereby reducing the risk factor;
   - change of occupancy and use.

3 *What information is required* by an inspector before conducting a Periodic Inspection and Testing?

   The installation information required by the inspector to conduct a *Periodic Inspection* is similar to the prerequisites for *Initial Verification*, such as diagrams, design criteria, electrical supply characteristics and earthing arrangements. If this information is not readily available, the person responsible for the installation should be able to supply it (see *BS 7671 Regulations 514.9 page 132*).

   Note: Where no previous documentations are available an investigation of the electrical installation should be undertaken before carrying out any inspecting and testing.

4 The *frequency of Periodic Inspection* will be determined by, for example:
   - the type of installation;
   - the type of equipment used and its operation;
   - the frequency of and quality of maintenance;
   - external influences;
   - any recommendation from previous reports.

   Note: If an installation is under effective management, where planned and preventive maintenance are continuously monitored by a skilled person competent in such work, and appropriate records are maintained, then a Periodic Inspection is *not* required.

5 *Setting Inspection and Testing samples*: The inspector will set the size of the installation sample to be inspected and tested; he or she will take into consideration the following:
   - approximate age and condition of the installation;
   - type and usage of the installation;
   - ambient environmental conditions;

- ongoing maintenance, if any;
- previous inspection/testing;
- the size of the installation;
- consultation with the installation owner;
- the quality of records.

Note: 100% Inspection and Testing in many installations is unrealistic, uneconomical and is possibly unachievable. A more realistic sample is 10%; however the size of the sampling should be made after a *walk about*, a review of previous records and the experience of the inspector.

Related topic questions

Q9. State the purpose of a Periodic Inspection and Testing.

   a) The installation is not in a satisfactory condition and can be continued to be used safely

   b) The installation is in a satisfactory condition and can be continued to be used safely

   c) It is necessary to confirm the installation has not deteriorated

   d) It is a legal necessity.

Q10. Under what circumstances would a formal Periodic Inspection of an electrical installation not be necessary?

   a) Where there is preventive maintenance and continuous monitoring by a skilled person

   b) Where preventive maintenance and continuous monitoring by a skilled person, competent in such work, is maintained

   c) Where preventive maintenance is not continuously monitored by a skilled person, competent in such work

   d) When a detailed examination of an installation can be carried out without dismantling, or any partial dismantling, taking place.

Related inspection and testing documents

The inspector needs to be familiar with *all* documents which are directly related to the Inspection and Testing procedure and be able to differentiate between Statutory and Non-Statutory documents.

The only Statutory document which is *directly* related to Inspection and Testing is the *Electricity at Work Regulations 1989*.

There are a number of Non-Statutory documents which are concerned with Inspection and Testing in one form or another; the most significant are:

- BS 7671 Requirements for Electrical Installations
- *IET On-Site-Guide*
- *IET Guidance Note 3 Inspection & Testing*

The candidate must be able to identify the correct title for this Non-Statutory document, for example BS 7671 is acceptable, whereas *18th Edition* is not.

Related topic questions

Q11. Which <u>Statutory</u> Regulation requires a given circuit or equipment to be <u>identified</u> before inspecting and testing procedure takes place?
  - a) 11
  - b) 12
  - c) 13
  - d) 14.

Q12. Under what circumstances would a duty holder invoke the Defence Regulation?
  - a) Only applies in criminal proceedings
  - b) Where reasonable steps have been taken
  - c) Due diligence has been exercised
  - d) Where criminal proceedings apply it provides a defence for the duty holder.

SUMMARY

There are a number of exam techniques which should be adopted to achieve the ultimate goal: exam success; however for some individuals the very thought of a theory exam can be extremely daunting. There are many who have not sat a formal exam since their school days. Regardless, the following approach has been devised for those individuals who may require some assistance in achieving the ultimate goal.

Typical exam techniques

- Read the entire question and the given answers carefully.
- Do not skim read the question or the given answers.
- Do not make assumptions, only select the answer which relates to the question.
- Do not spend valuable exam time pondering over a question where the answer is not immediately forthcoming or cannot be located easily; surprisingly the answer could be tucked away in another question or, equally, another question could simply jog the memory cells.

- The candidate must realise there are *no trick* questions; it is the candidate's competence, understanding, experience, skills and abilities, as an inspector, which are being tested.
- The candidate must recognise that he or she is being tested as an inspector *not* as a site electrician, installer or designer.
- It is crucial that the candidate (inspector) has a meticulous understanding and an in-depth knowledge of BS 7671 simply because the candidate's competence, understanding, skills, abilities and knowledge will be comprehensively tested by *City & Guilds* with their various and probing questions. Moreover, the theory exams are multi-choice and open book; therefore it is imperative that the candidate is fully conversant with BS 7671 and GN3; both can be referred to during the exam.

Answers and reasoning for related topics

Q1. Answer: b) Electricity at Work Regulations 1989.

Reasoning

The only *Statutory* document which is directly associated with an inspector's title and status, when conducting the *Inspection and Testing* of an electrical installation is the *Electricity at Work Regulations 1989*.

Note: *Electricity at Work 1989* is *not* an Act of Parliament but a set of *Regulations*.

Q2. Answer: a) Duty Holder.
Q3. Answer: c) A Competent Person.

Reasoning

The words to focus on in Q2 and Q3 are *title, status* and *law*:

- Law: the term "Law" refers to the Statutory document: *Electricity at Work Regulations 1989*.
- Title: an inspector must be *in control* of the installation when carrying out an Inspection and Testing procedure in order to prevent electrical danger and/or injury, therefore the inspector's title will be *duty holder*.
- Status: in order to prevent electrical danger and/or injury the inspector must be suitably qualified and *competent*.

(See *Memorandum of Guidance on the Electricity at Work Regulations 1989, Regulations 3 & 16 pages 11 & 39*.)

Q4. Answer: b) A skilled person (electrically).

Q5. Answer: c) A skilled person (electrically) competent in such work.

Q6. Answer: b) Skilled persons competent to verify that the requirements of the Standard have been met.

Reasoning

The term *legal status* is a direct reference to BS 7671, a Non-Statutory set of Regulations; however the same answer cannot be given for each question. The individuals who sign the EIC must be competent to verify/confirm the installation meets the requirements of BS 7671 (the Standard) for each element of the installation, that is: design, construction and Inspection and Testing (see *BS 7671 Part 2 Definition page 37, Part 6 Regulation 641.6 page 230, Regulations 644.5 page 236 and 653.5 page 238*).

Q7. Answer: c) The installation complies with the design and construction aspects of BS 7671.

Reasoning

The purpose of Initial Verification is to ensure the installation does comply with BS 7671 because these Regulations set the standard for the electrical installation's design and construction. If there is a supplementary question asking how this can be achieved, then a typical response could be: "the design and construction can be confirmed with the aid of an inspection & testing procedure *during and on completion* of the installation" (see *GN3 Initial Verification Section 2 paragraph 2.1 page 17*).

Q8. Answer: c) The value given by an enquiry to the DNO.

Reasoning

Initially this type of question could put a candidate in a slight quandary because there are only four sub-headings in the section covering *Supply Characteristics and Earthing Arrangement* and the question states that there are five items; however the question *does* refer to the supply characteristics, thereby directing the candidate to the sub-sections: *Nature of Supply Parameters* and *Supply Protective Device*.

If the sub-section, *Nature of Supply Parameters*, is reviewed, the nominal voltage (U/U_o [1]), the sub-script [1] adjacent to U/U_o requires the inspector to obtain the nominal voltage *by enquiry* (see BS 7671 *Part 3 Regulation 313.1 page 49 & Appendix 6 pages 463 & 474*).

Q9. Answer: b) The installation is in a satisfactory condition and can be continued to be used safely.

Reasoning

Possibly, an alternative question could be: why is it *necessary* to carry out a *Periodic Inspection?* However, the answer is not the same as for the *purpose* for conducting a Periodic Inspection and Testing.

There are a number of reasons why it is *necessary* to conduct a *Periodic Inspection*, for example: change of use of the premises (see *GN3 Section 3 pages 81–82*).

It is essential that the candidate carefully reads the question and can differentiate between the terminologies used: *purpose* and *necessary*, which are not the same (see *GN3 Periodic Inspection and Testing Section 3 paragraph 3.1*).

Q10. Answer: b) Where preventive maintenance and continuous monitoring by a skilled person, competent in such work, is maintained.

Reasoning

Although the word *necessary* is used, the question effectively relates to the words underlined. While there are a number of reasons *for* conducting a formal Periodic Inspection, when is a Periodic Inspection *not* required?

If an installation is continually monitored, and there is a system of preventive planned maintenance, any defects or problems will be detected; therefore the installation will be in a *satisfactory condition and can be continued to be used safely* (see *GN3 Section 3 paragraph 3.1 page 81 and BS7671 Regulation 652.2 page 237*).

Q11. Answer: b) 12

Reasoning

The question refers to a *Statutory* Regulation, and the only statutory document which relates to inspecting and testing is the *Electricity at Work Regulations 1989*. The Regulation which is directly associated with the first stage of isolation is *Regulation 12* (see *Memorandum of Guidance on the Electricity at Work Regulations 1989, Regulation 12 pages 31–33*).

Q12. Answer: d) Where criminal proceedings apply it provides a defence for the duty holder.

Reasoning

Regulation 29, referred to as the Defence Regulation, can apply when criminal proceedings are taken against a *duty holder*, however the *defence* for the duty holder is one of establishing proof that he/she took all reasonable steps and exercised all due diligence to avoid committing an offence; that is, following and applying the instructions given for a particular Regulation (see *Memorandum of Guidance on the Electricity at Work Regulations 1989, Regulations 29 page 41*).

2
Certification and reports

INTRODUCTION

It is extremely important for an organisation to maintain a portfolio of its electrical installation, in which all additions, minor works records, inspection and testing results and maintenance records are recorded throughout the lifespan of the installation. This action will assist in *detecting any deterioration, failings or defects within the installation at an early stage, thereby affording an electrically safe environment.*

There could also be another reason for maintaining a portfolio: it may be a stipulation for insurance cover.

The first stage in this process is the *Initial Verification*, during which the inspector carries out a full and detailed Inspection and Testing of an installation, primarily to ensure that *BS 7671 Wiring Regulations* fundamental principles have been met (the *Standard*), thereby confirming the installation has been designed and erected "*so as to provide for safety and proper functioning for the intended use*" and to verify that the designer's criteria has also been met.

The results of the inspector's Inspection and Testing undertakings will be recorded on an *Electrical Installation Certificate* (EIC) and the supporting documents, *Schedule(s) of Inspection* and *Schedule(s) of Test Results*. Thereafter the installation will be periodically checked and tested; the results should then be compared with previous Inspection and Testing results. This action will create a system of traceability and possibly the early detection of any deterioration in the installation's final circuits, which can be resolved quickly and efficiently without compromising the installation's safety.

Any additions or alterations, whether minor or otherwise, to the original installation must be recorded in order to maintain the concept of traceability and electrical safety. Therefore it is essential that the inspector fully understands his or her responsibilities in this process.

2.1 Documentation

All the Inspection and Testing documentation used in this Chapter is based on the *model forms* given in *Appendix 6 of BS 7671*. They may also be used or referred to in either of the C&G 2391 exams. They are:

1 *Electrical Installation Certificate*, which *must* be accompanied by:
   i) Schedule(s) of Inspection
   ii) Schedule(s) of Test Results.
2 *Minor Electrical Installation Works Certificate (Minor Works Certificate)*
3 *Electrical Installation Condition Report*, which must be accompanied by:
   i) *Condition Report Inspection Schedule* (for Domestic and Similar Premises with up to 100A Supply) or simply: *Schedule of Inspection*
   ii) Schedule(s) of Test Results
   iii) Guidance to Recipients.

2.2 *Electrical Installation Certificate (EIC)*

It is crucial for the candidate (inspector) to have a clear understanding of the individual elements of the EIC, not just for C&G exams but *for practical inspecting and testing applications*. The latter is essential because C&G questions will be geared in such a manner as to ensure the candidate fully comprehends all aspects of the *Electrical Installation Certificate*; especially since C&G's Chief Examiner has highlighted the poor response candidates have given to questions relating to this document under the previous exam format.

In the defence of many skilled and competent electricians, it is highly probable that the act of completing an EIC is just routine, therefore the minutiae of the EIC is not always fully absorbed. Any breaches in their intimate knowledge of the EIC are only revealed when questioned on the document, which should not impeach the professionalism and competence of the skilled electricians during the Inspection and Testing process. Consequently, the candidate *must* be familiar with the finer details of the EIC, for example:

Questions

1 What applications can the EIC be used for?
2 Who will determine when the next Inspection and Testing should be undertaken (first Periodic)?
3 Who signs the EIC?

4   What is the status of the signatories?

5   What documents must accompany an EIC to confirm its validity?

6   What general characteristics must be made available to the inspector?

7   What information must be made available to the inspector?

8   Who determines the nominal voltage and frequency?

9   What are the titles of documents used (correct terminology is essential)?

10   What information needs to be recorded for the overcurrent protective device(s) at the origin of the installation?

Responses

1   The applications for the EIC are:
     i)   New installation
     ii) Addition to an existing installation
     iii) Alteration to an existing installation.

2   The next Inspection and Testing is determined by the designer(s) of the installation.

3   The signatories on the EIC are:
     i)   designer(s)
     ii) constructor/installer
     iii) inspector
     iv) If a single signature document is used, the person signing the EIC will be responsible for all three elements: design, construction, inspecting and testing.

4   They are *all* skilled persons competent in such work.

5   Schedule(s) of Inspection and Schedule(s) of Test Results must accompany the EIC to ensure its validity (the correct terminology *must* be used).

6   The *Supply Characteristics and Earthing Arrangements,* which should be made available to the inspector, as required by BS 7671 in Part 3 Sections 312 and 313 are:
     i)   Earthing arrangement
     ii) Number and type of live conductors
     iii) Nature of Supply Parameter
     iv) Supply Protective Device.

     Although *Maximum Demand* is also given in Part 3 Section 311 of BS 7671, it is referred to under the general heading of *Particulars of Installation Referred to in the Certificate,* and should not be confused with those items given above in 6 (i)–(iv).

7   The information which should be made available to the inspector is given in Chapter 51 Regulation 514.9.1 *Diagrams & Documentation*, which is:

i)   Type and composition of each circuit

ii)   Method used for basic and fault protection

iii) Identification of each device performing the function of protection, isolation and switching plus location

iv)   Circuit or equipment vulnerable to electrical tests (i.e. insulation resistance tests).

(Note the difference between items 6 and 7.)

8   The nominal voltage and frequency, in the element *Nature of Supply Parameters*, has a subscript (1) indicating that these nominal values could only be determined by *enquiry* and *not* 230/400 volts and 50Hz. Although *Appendix 2* of BS 7671 does give the nominal voltage as 230 volts + 10/–6 %, furthermore there is the correction factor (C_{min}) of 0.95 used when calculating an earth fault loop impedance (Z_s) value, nevertheless the requirements of the EIC take precedence.

9   The correct terminology must be used for *all* the documents which are finally handed to the "person ordering the work" on the satisfactory completion of the Inspection and Testing. These documents are: *Electrical Installation Certificate*, *Schedule of Inspection* and *Schedule of Test Results*. The correct wording for each document is crucial; any other terminology or wording is not acceptable.

10   The information which should be recorded in the element *Supply Protective Device*, for the overcurrent protective device(s) at the origin of the installation, is:

i)   *Type – the type of protective fuse used, e.g. BS 88 or BS 1361, the latter has been re-configured as BS 88-3*

ii)   *Rated Current – the current rating, in amperes, of the protective device, e.g. 60A, 80A or 100A.*

The protective device(s) is the property and responsibility of the energy supplier's DNO (Distribution Network Operator – e.g. N-Power, EDF, EON), therefore if the information is *not* readily available it should be obtained from the relevant DNO.

The inspector is advised *not* to insert the word *limit* for this element (see Regulation 313.1 Item (vi) of BS 7671).

Related topic questions

Q1. What <u>documents</u> are handed to a client on the completion of replacing a consumer unit?

a) *Electrical Installation Certificate, Schedule of Inspection* and *Schedule of Test Results*

b) *Minor Electrical Installation Works Certificate, Inspection and Tests Schedules*

c) *Electrical Installation Certificate, Schedule of Inspection, Schedule of Test Results & Guidance for Recipient*

d) *Minor Electrical Installation Works Certificate.*

Answer: c) *Electrical Installation Certificate, Schedule of Inspection, Schedule of Test Results & Guidance for Recipient.*

Reasoning

Regardless of whether the replacement of a consumer unit is *like-for-like*, an *Electrical Installation Certificate* must be completed, <u>not</u> a *Minor Electrical Installation Works Certificate*. An *Electrical Installation Certificate* must always be accompanied by the two Schedules; *the correct terminology must be used*, whereas the Minor Works Certificate is a single document.

Q2. What information is recorded on an EIC with regards to the <u>Distribution Network Operator's (DNO) protective device</u>?

a) This is not the inspector's responsibility

b) It is the responsibility of the installation constructor

c) BS (ES), Type and Rated Current

d) Fuse Type, Current Rating and BS Number.

Answer: c) BS (ES), Type and Rated Current.

Reasoning

The key words in this question are <u>Distribution Network Operator</u> and <u>protective device</u>; the former is the energy supplier whereas the latter is self-explanatory. The only section on the *Electrical Installation Certificate* which links the two is given in the *Supply Protective Device* and there are three elements, which are, in the sequence given on the EIC: BS (EN), Type and Rated Current, for example BS 88, Series 7 and 80A.

2.3 *Schedule of inspection*

The *Schedule of Inspection* is an extremely comprehensive document inasmuch as each item inspected must comply with its relevant *Regulation* with the exception of *Section 1*, which relates to the *Distributor's Supply Intake Equipment*.

Although the six items in *Section 1* are effectively under the aegis of the related DNO they should not be neglected either in practice or during the relevant City & Guilds exam.

The inspector will be inspecting the *condition* of the items given in *Section 1*, which are:

- Service cable
- Service head
- Distributor's earthing
- Distributor/consumer meter tails
- Metering equipment
- Isolator, where fitted.

If there is a problem, the relevant DNO needs to be informed.

The candidate should not lose sight of his/her responsibilities when completing the *Schedule of Inspection*. The objective is to *inspect*, an observation exercise, to check and confirm whether, for example: cables and protective devices are actually in place, equipment and enclosures are not damaged and comply with standards. To achieve these objectives the inspector will need to use a number of his/her senses. They are:

1 Sight
2 Touch
3 Smell
4 Hearing.

The *Schedule of Inspection for New Installations* is divided into eleven Sections. They are:

1 Distributor's supply intake equipment
2 Parallel or switched alternative sources of supply
3 Automatic Disconnection of Supply (ADS)
4 Basic protection
5 Additional protection
6 Other methods of protection
7 Consumer unit(s)/distribution board(s)
8 Circuits

9   Current using equipment (permanently connected)
10 Location(s) containing a bath or shower (Section 701)
11 Other Part 7 Special Installations or Location

1 Distributor's supply intake equipment

The six items in this Section are visually inspected, primarily for their *condition*.

2 Parallel or switched alternative sources of supply

The inspector should confirm that there are *adequate* arrangements for operating alternative energy sources with the public supply; insofar as there are means to switch or parallel these different types of electrical energy with public supply.

3 Automatic disconnection of supply (ADS)

ADS depends upon earthing and bonding for its successful operation; therefore the inspector must confirm the *presence and adequacy* of the installation's *earthing and protective bonding arrangements*, in the respect:

- **Presence of Earthing Conductor** – between the consumer's main earth terminal (MET) to the supplier's protective arrangement; which are:
  i)   TN-S system uses the steel wired armour (SWA) of the supplier's cable
  ii) TN-C-S the earth cable is connected to the supplier's neutral
  iii) TT uses the main body of earth, with the earth conductor connected to an earthing rod or earth electrode driven into the main body of earth.

  The inspector's inspection responsibilities are:
  i)   To confirm the earth conductor is correctly connected and secure
  ii) To confirm, where the MET is external to the distribution board or consumer unit, there is a durable label marked: *Safety Electrical Connection Do Not Remove* is attached
  iii) The durable label (as ii) is attached to the TT's earth electrodes
  iv) To ensure the ratio between the cross sectional area (csa) of the earth conductor and supplier's line conductor meets the requirements of BS7671 *Table 54.7.*
- **Presence of circuit protective conductors (cpc)** – should be connected to each point and accessory in the final circuit to the MET, except *lampholders*, which have no "exposed-conductive-parts and suspended from such a point".

- **Presence of the Main Equipotential Bonding Conductors** – All the extraneous *metalwork* of the incoming services, such as water, gas and heating oil, plus any exposed and structural steelwork, are connected to a main earthing terminal (MET). This requirement does not apply where: "Metallic pipes entering the building having an insulation section at their point of entry need not be connected to the protective equipotential bonding" (BS7671 Regulation 411.3.1.2 page 58).

  Initially the inspector will need to confirm, by inspection:

  i)   The bonding conductors are correctly connected and secure
  ii) A durable label marked *Safety Electrical Connection Do Not Remove* is securely attached to the incoming services and steelwork connections
  iii) Where the MET is external to the distribution board or consumer unit the durable label is also securely attached
  iv) To ensure the ratio between the cross sectional area (csa) of the bonding conductor and supplier's neutral conductor meets the requirements of *BS 7671 Table 54.8*.

- **Presence of Supplementary Equipotential Bonding Conductors** – Confirm all *exposed conductive parts and extraneous conductive parts* are bonded at the MET in order to maintain *both types of conductive parts* at the same potential, to ensure a potential difference does not occur under fault conditions.

  Note: Supplementary equipotential bonding is also used as additional protection.

4 Basic protection

In previous editions of BS 7671, *basic protection* was referred to as *direct contact*, which probably gave a more meaningful perception of the type of protection expected by many electricians. Perhaps a more blunt approach should be taken: *can you touch an exposed live conductor?* Or: *Is any part of the insulated live conductor exposed to touch?* The answer is either yes or no to either type of question, which is the reason for conducting the *insulation of live parts* exercise.

- **Barriers or Enclosures** – Generally *barriers* are there to protect an individual from inadvertently touching or coming into contact with live parts behind the barrier, whereas an *enclosure* is the outer casing of electrical equipment with *live* parts inside it.

  Irrespective of whether it is a barrier or an enclosure which is being inspected, the inspector should be looking for:

  - Any signs of damage or defects
  - Does the top horizontal surface comply with *Regulations 416.2.2?*
  - Does the enclosure or barrier comply with *Regulations 416.2.1?*

Regulations 416.2.2 simply states: the horizontal top surface of either a barrier or an enclosure, which is readily accessible, shall provide a degree of protection of at least IP 4X or IP XXD. Both codes will provide protection against the ingress *of wire or strips and solid objects equal or greater than 1mm and 1mm²* *respectively*; whereas the latter relates to additional penetration protection against contact with live parts to a depth of 100 mm.

Regulations 416.2.1 states: live parts shall be inside enclosures or behind barriers providing at least a degree of protection of IP2X or IPXXB. Both codes relate to the protection against the ingress of solid particles which are equal to or greater than 12.5 mm², however the latter provides supplementary penetration protection for human body parts, such as a *finger* coming in contact with a live part at a depth not exceeding 80 mm.

Note: IP2X and IPXXB do *not* apply where large openings are necessary to allow the correct functioning of given equipment.

5 Additional protection

This form of protection is applicable to:

i) Residual Current Device (RCD)
ii) Supplementary Bonding
iii) Where RCDs are used for additional protection, they should not exceed 30mA, or be used as the sole means of protection.

RCD protection extends to all types of electrical installations: domestic, commercial and industrial. This protection covers socket outlets with a current rating not exceeding 32A regardless of their location; including those supplying mobile equipment which are used outdoors

This additional protection also applies to AC final circuits supplying luminaires in domestic (household) premises.

Special RCD consideration

If a designer, for whatever reason, decides to exclude RDC protection for socket outlets a *documented risk* assessment must be completed, stating the reason(s) for the exclusion, and attached it to the *Electrical Installation Certificate* and, if applicable to the *Minor Electrical Installation Works Certificate (Minor Works Certificate)*.

Note: This exclusion does <u>not</u> apply to socket outlets within domestic dwellings.

i) Supplementary Equipotential Bonding is normally checked during the inspection of *Automatic Disconnection of Supply* in Section 3, however the inspector should be aware *Supplementary Equipotential Bonding* is also an *additional form of protection*.

6 Other methods of protection

Both basic and fault protection can be provided by:

- SELV – separated extra low voltage
- PELV – protective extra low voltage
- Double or reinforced insulation
- Electrical separation for one item of equipment.

Where appropriate, the presence and effectiveness of this type of protection shall be inspected and tested.

7 Distribution doards (DB) and consumer units (CU)

The inspection and checking of DBs or CUs can be routine; unfortunately this can prove to be a major stumbling block when questioned on the subject simply because it *is just* routine. Possibly the most constructive advice which can be given to overcome this problem is for the candidate to mentally visualise a DB or CU and then ask the following questions.

When inspecting the enclosure what would the inspector be checking?

i) Initially, what is the condition of the enclosure? Is there any damage?

ii) Are there any warning notices? If so, what would they be? Are any of the following warning notices necessary?

- Voltage – nominal voltage exceeds 230 volts to earth
- RCD – six monthly test notice
- Non-standard colours
- Alternative supplies
- Periodic Inspection and Testing.

iii) Are the cables entering the DB suitable protected?

iv) Is the enclosure IP and fire rating compliant?

What would the inspector be expected to see when the DB's door is opened?

i) Are there any diagrams, charts or schedules in a plastic envelope attached to the DB's door?

ii) Are the final circuits clearly and correctly identified?

iii) Is the enclosure suitably IP rated?

iv) Is there an isolation warning notice present?

When the enclosure is removed what would the inspector see inside the DB and what would he/she be expected to check?

i) Is there a barrier?

ii) Is it suitably IP rated?

iii) Is there any damage to the conductor's insulation?

iv) If the DB is of metallic construction, is it suitably bonded?

v) Are there any SWA protected cables? If so, is the SWA bonded with the appropriate earthed banjo?

vi) Are the conductors secured in their appropriate connections?

vii) Are all the conductors: lines, neutrals and cpcs in the correct numerical sequence?

viii) Are the conductors current rating (I_z) compatible with their associated protective devices?

ix) If RCDs are fitted do they meet BS 7671 standards?

x) If circuit breakers are fitted are they mechanically operable (functional check)?

xi) Is the main isolation switch mechanically sound (functional check)?

8 Circuits

Section 8 effectively formalises the inspection process itemised in IET's Guidance Note 3 *Inspection & Testing*, in which very useful comments and guidance are given.

Although the requirements of Section 8 are reasonably routine in terms of inspecting and checking, the inspector is now formally required to confirm, for example, whether the cables have been examined for any defects, they have adequate current carrying capacity and are colour coded correctly.

Furthermore, the inspector will need to visually inspect different types of wiring systems and their enclosures, whether the wiring systems are either:

i) metallic conduit or trunking systems, or

ii) pvc conduit or trunking systems.

The enclosures should be examined to ensure:

i) they are free from burrs

ii) they are IP and fire rated

iii) they are secured and erected correctly with the appropriate fixing

iv) they are free from contaminates

v) continuity of metallic enclosures

vi) compatibility with external influences and location.

In addition to the inspection of cables and wiring systems, the inspector needs to check:

i) additional protection

ii) isolation and switching.

i) Provision of additional protection by RCD not exceeding 30mA

(see BS 7671 *Regulation 411.3.3:* page 59 *Additional Protection* page 73).

Additional protection provided by 30mA RCD for cables in concealed in walls (see BS 7671 *Regulation 522.6.202 page 139).*

ii) Isolation and switching

The correct location of an appropriate device for either *isolation* and/or *switching* is the responsibility of the designer who will specify the type, its function and location.

There are four functional categories. They are:

• isolation

• switching for mechanical maintenance

• emergency switching

• functional switching.

The inspector's responsibilities are to ensure the equipment is:

- correctly located
- suitably identified
- readily accessible

and

- capable of being locked off (where appropriate), and that
- warning notices are in place, and
- functional checks have been made.

SELECTION OF APPROPRIATE FUNCTIONAL SWITCHING DEVICES

Functional switching is defined as *an operation intended to switch "on" or "off" or vary the supply of electrical energy to all or part of an installation for normal operating purposes.*

A review of the designer's criteria will disclose the location, functional operation and the type of switching devices, which could include, for example:

- switchgear and controlgear assemblies
- drives
- controls
- interlocks.

The inspector should carry out a functional check (*does the device work?*) to ensure the equipment is correctly erected, adjusted and installed in accordance with the relevant requirements of BS 7671 Regulations.

Where an RCD is fitted for fault protection and/or additional protection, its functional effectiveness must be verified with the aid of the test button incorporated in the device.

Note: This functional test verifies the *mechanical operation* of the RCD, not the electrical operation.

9 Current using equipment (permanently connected)

During the Initial Verification process, the inspector will examine all fixed current using equipment to ensure it is:

- free from enclosure damage
- suitable for the environment
- correctly IP and fire rated

- fitted securely
- suitably undervoltage protected
- suitably overload protected.

Also, that downlighters (recessed luminaires) are fitted with the correct type of lamps and they have adequate ventilation for heat dissipation.

10 Special installation or location (Part 7)

There are twenty one *Special Installations or Locations* in Part 7 and each *Special Installation or Location* will have its own unique requirements. The following are just a few examples highlighting some of the more salient elements.

Location containing a bath or shower

- All final circuits within the location must be protected with 30mA RCD.
- All external final circuits, to the location but passing through Zones 1 and 2 must be protected with 30mA RCD.
- Ingress protection for Zone 0 is IPX7: the immersion code.
- For Zones 1 and 2, it is IPX4: the splash code.
- However, if electrical equipment in Zone 2 is subjected to water jets, the IP Code is IPX5.
- Supplementary bonding is *not* required in the location provided:
  - i) The disconnection times for the location's protective devices comply with those given in *Regulation 411.3.2 and Table 41.1.*
  - ii) All final circuits are RCD protected.
  - iii) All extraneous-conductive parts are effectively connected to the protective bonding, e.g. incoming metallic water and gas comply with *Regulation 411.3.2.*

Swimming pools and other basins

All electrical equipment in the following Zones should have the minimum protection of:

- IPX8 in Zone 0, which is the water-filled swimming pool where electrical equipment, and humans, can be fully submersed.
- IPX4 in Zone 1 or IPX5 if water jets are used in this Zone for cleaning purposes.
- IPX2 in Zone 2 provided the equipment is indoors, IPX4 where the equipment is outdoors and IPX5 if water jets are used.

- Fountains only have two Zones and they are Zones 0 and 1; there is no Zone 2.
- Where PME earthing is used, the earth mat or earth electrode should have a suitably low resistance of 20Ω or less.

Construction and demolition site installations

- Final circuits supplying socket outlets with a rated current exceeding 32 A, Regulation 411.3.2.5 page 59, is not applicable.
- Socket outlets exceeding 32A shall be protected with an RCD not exceeding 500mA.
- RCDs not exceeding 500mA are only referred to in this *special location* in BS 7671.
- PME is not to be used unless all extraneous-conductive-parts are reliably connected to the installation main earth terminal (MET).

Agriculture and horticulture premises

- 30mA RCD protection is required for final circuits supplying socket outlets not exceeding 32A.
- 100mA RCD protection is required for final circuits socket outlets exceeding 32A.
- 300mA RCD protection is required for other circuits and for fire protection.
- Depth of buried cables
  i) In areas around agricultural premises, where there is vehicular movement, to at least 0.6 m with mechanical protection.
  ii) In arable and cultivated land, to at least 1 m with no mechanical protection stipulated.
  iii) Overhead, self-supporting cables to a height of at least 6 m.
- IP 44 coding.
- Conduit wiring systems to with stand an impact of 5J.
- TN-C earthing is not to be used.
- PME earthing is not to be used where livestock are located on concrete floors unless a metal re-enforcing grid is included.

Electrical installations in caravan/camping parks and similar locations

- Overhead cables, where there is vehicular movement, not less than 6 m, in other areas 3.5 m.

- PME earthing is not permitted where there is metalwork in leisure accommodation, including caravans.

Related topic questions

Q3. A four core steel-wire armoured (SWA) cable is terminated in a distribution board with one core being used as the cpc; what human senses are appropriate to confirm whether the SWA gland and the conductor's terminations meet the current Standard during the inspection process?

a) Touch, smell and heat
b) Sight and touch
c) Sight, touch and smell
d) Hearing, sight, touch and smell.

Answer: b) Sight and touch.

Reasoning

The question clearly identifies the type of wiring system and what is required; that is, are the gland and the conductors terminated correctly? Therefore the only human senses that are appropriate are touch and sight, more so because the inspection is a *dead test*.

2.4 Minor electrical installation works certificate (minor works certificate)

This particular *Certificate* is possibly the most neglected of all the Inspection and Testing documents, probably because little thought is given to its applications, which are:

- Additions – additional socket outlets or lighting points to *existing* circuits.
- Alterations – the relocation of light switches or the replacement of equipment but it does *not* extend to new circuits or the *replacement of distribution boards or consumer units.*

The inspector (candidate) must be familiar with each *Part* of the *Certificate*, and recognise the significance of this document in the respect that it is a *Certificate* and not a *Report*; consequently the conditions that apply to the *Electrical Installation Certificate* also apply to the *Minor Works Certificate.*

Parts 1–5

The data/details for the *Minor Works Certificate* are extremely similar to that of an *Electrical Installation Certificate*.

The Five Parts of the Certificate are:

Part 1: Description of the Minor Works
The following details are required:
- name of the client,
- the location/address of the work to be undertaken,
- the description of the work, and
- the date the work was completed.

If there are any departures *from Regulations 120.3, 133.3 and 133.5* they must be recorded.

Comments on the existing installation are also required.

There is also a requirement for a *Risk Assessment* document to accompany the *Certificate* where additional protection for socket outlets is not provided as required by *Regulation 411.3.3* (Also see Appendix 2 Item 11 page 360)

Part 2: Presence and Adequacy of Installation Earthing and Bonding Arrangement
The following are required:
- System's earthing arrangement: TN-S, TN-C-S & TT
- Earth Fault Loop Impedance at DB for the final circuit
- Presence of adequate main protective conductor: Earthing Conductor and Equipotential Bonding

Part 3: Circuit Details to include:
- DB Number, location and type
- Circuit Number and description
- Circuit's overcurrent protective device: BS(EN), Type and Rating
- Conductor size: Live and cpc

Part 4: Test Results for the Circuit altered or Extended
- Protective conductor continuity: $R_1 + R_2$ or R_2
- Continuity of Ring Final Circuit Conductors: L/L, N/N & Live-Earth
- Polarity Satisfactory
- Maximum Zs
- RCD operation: Rated residual operation current, Disconnection time and Satisfactory test button operation

Part 5: Declaration – the person signing the Certificate is confirming the safety of the installation is not impaired and " *to the best of their knowledge and belief, at the time of my/our inspection, complied with BS 7671 except as detailed in Part 1*"

Information to be given to the client

- The *original Certificate* should be given to *the person ordering the work.*
- If the person ordering the work is *not* the owner, the *Certificate* or a copy should be given to the owner.
- A *duplicate copy* of the *Certificate* should be *retained by the contractor.*
- The *Certificate* is only valid for the minor work carried out on individual circuits.
- A document giving guidance to the recipient of the minor works.

Related topic questions

Q4. State the information which must be recorded on the <u>*Minor Electrical Installation Works Certificate*</u>, which relates to Distribution Network Operator's supply.
- a) Supply fuse and current rating
- b) System earthing arrangement
- c) System earthing arrangement and method of fault protection
- d) The protective device for the modified circuit.

Answer: b) System earthing arrangement.

Reasoning

There are three Items in Part 2 of the *Minor Electrical Installation Works Certificate* which relate to the *Installation* and only one of these items is linked directly to the supply input; that is, the *System earthing arrangement*, which can be either TN-C-S, TN-S or TT .

Parts 3 & 4 are directed towards the modified circuit in terms of its *method of fault protection and protective device.*

Furthermore, there are a series of electrical tests which must be carried out on the modified circuit and recorded.

One of these tests requires the inspector to check whether the system's earth continuity is satisfactory. The problem the inspector may encounter when conducting this test is: how will he/she know whether the earth continuity is satisfactory unless the system's earthing arrangement is known?

If this type of logic is applied it could help a candidate respond successfully to an exam question.

2.5 *Electrical installation condition report (EICR)*

The *Electrical Installation Condition Report* (EICR) is a comprehensive document, which the candidate (inspector) needs to be extremely familiar with. For examples, see the following questions.

Questions

1 Who should the completed documents be handed to?
2 What applications can the EICR be used for?
3 What installation details need to be recorded?
4 Who should the extent and limitations of the Inspection and Testing be agreed with?
5 Is there a need to inspect the electrical equipment in the accessible roof space of a house?
6 Under what circumstances will an installation be declared unsatisfactory?
7 Who will determine when the next Inspection and Testing should be undertaken?
8 Who signs the EICR?
9 What is the status of the signatories?
10 What documents must accompany an EICR to confirm its validity?
11 What general characteristics must be made available to the inspector?
12 What information must be made available to the inspector?
13 Who determines the nominal voltage and frequency?
14 What are the titles of the documents used (correct terminology is essential)?
15 What information needs to be recorded for the overcurrent protective device(s) at the origin of the installation?

The responses to these questions are effectively given in the eight alphabetical Sections (A–K) of the EICR. Each Section needs to be read carefully and fully understood by the candidate.

Responses to questions 1 to 11

1 **Section A** – *Details of the Client/Person Ordering the Report* and the completed documents will be issued to this person.
2 **Section B** – *Reason for Producing the Report* – insurance, change of ownership etc.

3 Section C –*Details of the Installation which is Subject to this Report* – the *occupier* could be a tenant and the *address* could be different from the actual owner of the installation. The candidate must be familiar, and aware, of the *boxes* which need to be ticked in this Section.

4 Section D – *Extent and Limitation of Inspection and Testing*: The *extent and limitations* need to be established *with the client*. *Limitations* (lim) are *not* permitted unless there are *operational reasons* or they have been *agreed with a named person*, supported with the *reasons* for the limitations. All these details must be clearly recorded.

5 The explanatory notes, given in Section D, states the items which can or cannot be expected to be tested.

6 Section E – *Summary of the Condition of the Installation*: The condition of the installation is either *satisfactory* or *unsatisfactory*. If the installation is categorised as either *C1, C2 or FI* it is unsatisfactory.

7 Section F – *Recommendation*: This is an informative Section and recommends the action(s) the client should take if the observation Codes C1, C2 or FI are allocated and *recommends a date for a re-test once remedial work has been completed*. The client is also advised to give *due considerations* if the observation Code C3, *improvement recommended*, is given.

   Note: It is the *inspector* who will recommend the date of the next Periodic Inspection and Testing

8 Section G – *Declaration* – Status of the signatories: A skilled person who is competent in such work. The person who carried out the Inspection and Testing and the person who authorised the Report to allow it to be issued

9 When this Section is signed by the inspector it effectively validates the Inspection and Testing of the installation; which can be substantiated by a suitable qualified person who will *authorise the issue of the Report*. This action is generally associated with large organisations that may employ a *qualified supervisor*. Conversely, the inspector is not precluded from signing both parts of the declaration simply because he or she will have an "*appropriate level of education and gained sufficient experience and knowledge to be fully conversant with the aspects required to carry out such an important inspection*".

10 Section H – *Schedule(s)*: The *Electrical Installation Condition Report* is *not* valid unless it is accompanied by *Schedule(s) of Inspection* and *Schedule(s) of Test Results*. The number of Schedules must be recorded in this Section.

11 Sections I & J – *Supply Characteristics and Earthing Arrangements* and *Particulars of Installation Referred to in the Report* are similar to those details given on the *Electrical Installation Certificate (EIC)*, namely: Earthing Arrangements, Number and Types of Live Conductors, Nature of Supply Parameters and Supply Protective Device.

Section K – *Observations*: The condition of the final circuits and whether they meet the requirements of BS 7671 are acknowledged in the box marked "*No remedial action is required*". If, however, any of the final circuits fail to comply with the required standard then an appropriate Code of C1, C2, FI or C3 will be noted in the box marked "*The following observations are made*" with a tick, supported with a full explanation of the problem.

Responses to questions 12 to 15

Q12: What information must be made available to the inspector?

The *Supply Characteristics and Earthing Arrangements* as given in Section I of the EICR should be made available to the inspector before carrying out the Inspection and Testing procedure. They are:

- Earthing arrangements
- Number and Type of Live Conductors
- Nature of Supply Parameters
- Supply Protective Device.

(Also see Section 2.3 of IET Guidance Note 3. and BS7671 Chapters 311, 312 & 313 pages 45–49)

Moreover, relevant information that will assist the inspector to *carry out safe* Inspection and Testing is highlighted in BS 7671 Regulation 514.9.1 *Diagrams and Documentation*, which includes:

i) Type and composition of each circuit.
ii) Method used for basic and fault protection.
iii) Identification of each device performing the function of protection, isolation and switching plus location.
iv) Circuit or equipment vulnerable to electrical tests (i.e. insulation resistance tests).

Q13. Who determines the nominal voltage and frequency?

The nominal voltage and frequency, in the element *Nature of Supply Parameters*, has a subscript (1) indicating these nominal values could only be determined by *enquiry*.

Q14. What is the title of the documents used (correct terminology is essential)?

   i)   *Electrical Installation Condition Report*
   ii) *Condition Report Inspection Schedule*
   iii) *Schedule of Test Results.*

Q15. What information needs to be recorded for the overcurrent protective device(s) at the origin of the installation?

Type – the type of protective fuse used, e.g. BS 88 or BS 1361 – the latter has been re-configured as BS 88-3, and *Rated Current* – the current rating, in amperes, of the protective device, e.g. 60A, 80A or 100A.

Note: With the exception of Q14, the answers given to questions 12–15 mirror similar questions posed for the *Electrical Installation Certificate*.

Coding explanation

- C1 (**Danger present**) – *the safety of those using the installation is at risk*, and it is recommended that a skilled person competent in electrical installation work undertakes the necessary *remedial work immediately*.
- C2 (**Potentially dangerous**) – *the safety of those using the installation may be at risk* and it is recommended that a skilled person competent in electrical installation work undertakes the necessary *remedial work as a matter of urgency*.
- C3 (**Improvement recommended**) – normally attributed to older installations designed prior to BS 7671:2008 and do not meet the current standards.
- FI (**Further investigation**) – the inspection has revealed an apparent deficiency. Further investigations should be carried-out without delay to determine the problem.

Note: i) If either C1, C2 or FI codes are given, the installation is considered to be *unsatisfactory*;

ii) If *more than one Observation Code* can be attributed to a final circuit, the more serious Code should be allocated.

2.6 *Condition report inspection schedule* (for domestic and similar premises with up to 100A supply)

The current *Schedule of Inspection* has undergone radical and significant changes inasmuch as every item inspected has a *Regulation* reference to which the inspector should cross reference as necessary. This concept reduces the possibility of error appreciably and ultimately gives an element of accountability.

There is, however, one exception: *Section 1.0 Distributor's Supply Intake Equipment*, which has no BS 7671 *Regulation* reference. Nevertheless it is the responsibility of the inspector to *review the condition* of the respective items and record, if necessary, an appropriate Code of C1, C2, C3 or FI.

Note: The ultimate responsibility for reporting any defectives with the *Distributor's Supply Intake Equipment* resets with the person ordering the work.

The items to be inspected on the *Condition Report Inspection Schedule* are similar to those given on the *Schedule of Inspection for New Installation Work*: although the inspection process will be similar, the outcomes can be different.

When the inspector has completed his/her Inspection and Testing of an installation, the appropriate documentation can be issued to the client simply because the inspector will be *reporting* on the *condition* of the installation, whereas no documents can be issued for an *Initial Verification* until the installation meets the standard required by BS 7671.

2.7 *Schedule of test results*

The *Schedule of Test Results* is reasonably straightforward. There are clear headings denoting what is to be recorded, there are group headings where more than one item of information is required, plus each column is also in numerical order.

Items 3–11, on the Schedule, give *Circuit Details* relating to the *Protective Device* used and the cross sectional area (csa) of the live and cpc conductors used; whereas Items 12–23 relate to instrument Test Results.

Explanation for some of the columns

Column Number 6   Breaking Capacity (kA) Refers to the protective device's Icn rating shown as 6000 on the circuit breaker but recorded in kA

Column Number 7   The nominal value of the RCD is recorded

Column Number 9 Reference Method The *type of wiring method* given in Appendix 4.

Column Number 17 The insulation resistance test voltage is recoded for each final circuit, ie 250, 500 or 1000 DC voltages

Column Number 24 Used to record the mechanical function of an AFDD

The following values must also be recorded on the *Schedule of Test Results*:

- **DB reference number,** which is generally applicable to large installations where there could be more than one Distribution Board.
- **Location of the DB.**
- **Zs at DB(Ω).** This ohmic value will be recorded for each DB or consumer unit (CU) inspected and tested. If there is only one DB or one CU, the ohmic value recorded will be for the external earth fault loop impedance Ze. This value will also be recorded on the *Electrical Installation Certificate* and *Electrical Installation Condition Report*.
- **Ipf at DB (kA).** The prospective short circuit current will be recorded for each DB or CU inspected and tested. However, if the DB is at the origin of the installation the value will also be recorded on the *Electrical Installation Certificate* and *Electrical Installation Condition Report*.
- **Correct supply polarity confirmed.** The polarity must be confirmed at each DB or CU being inspected and tested. It is also recorded on the *Electrical Installation Certificate, Electrical Installation Condition Report* and the *Minor Electrical Installation Works Certificate*.
- **Phase sequence confirmed.** This will only be applicable for three phase systems.

Related topic questions

Q5. On completion of a Periodic Inspection what documents are handed to the client?

a) *Electrical Installation Certificate, Schedule of Inspection* and *Schedule of Test Results*

b) *Electrical Installation Condition Report, Schedule of Inspection* and *Schedule of Test Results*

c) *Electrical Installation Certificate, Schedule of Inspection, Schedule of Test Results and Guidance for Recipient*

d) *Electrical Installation Condition Report, Schedule of Inspection, Schedule of Test Results* and *Condition Report Guidance for Recipient.*

Answer: d) *Electrical Installation Condition Report, Schedule of Inspection, Schedule of Test Results* **and** *Condition Report Guidance for Recipient.*

Reasoning

Answer b) is probably the most obvious answer simply because the EICR is not valid unless appropriate, correctly completed, *Schedules* are attached to the Report. However, a Condition Report Guidance for Recipient document must also be attached to the EICR. Unfortunately, any reference to the Guidance document is generally neglected.

(See *BS 7671 Appendix 6 page 476.*)

Q6. During a Periodic Inspection an apparent deficiency was made but could not be fully identified owing to the extent and limitation imposed in Section D of the EICR. What observation code and Summary of the Condition of the Installation should the inspector record in Sections E & K of the Report?
a) C1 Unsatisfactory
b) C2 Unsatisfactory
c) FI Unsatisfactory
d) C3 Satisfactory.

Answer: c) FI Unsatisfactory.

Reasoning

A Further Investigation (FI) observation code must be recorded because the apparent deficiency could not be identified, for the reasons given. Subsequent detailed inspection could reveal a problem which could lead to the generation of either a C1 or C2 observation code, thereby rendering the installation unsatisfactory.

(See *BS 7671 Appendix 6 item 9 page 475.*)

Prohibited or restricted earthing systems in special installations or locations

Protective Multiple Earthing (PME) is prohibited or subject to specified restrictions in the following installations:

- **Swimming Pools and Other Basins** may be permitted provided there is an earth mat or earth electrode with a low resistance, 20Ω or less is recommended, and is connected to protective equipotential bonding.
- **Construction and Demolition** sites may be used provided all extraneous-conductive-parts are reliably connected to the main earth terminal (MET).
- **Agricultural and Horticultural** premises, where there are locations with concrete floors and no metal reinforcement grid system intended to be used for livestock.

- **Caravan/Camping Parks,** including the metalwork of leisure accommodation vehicles and caravans, may be used where there are permanent buildings on the sites.
- **Marinas** may be used where there are permanent buildings on the sites.
- **Exhibitions Shows and Stands** can be used provided the installation is continuously supervised by a skilled person or an instructed person and suitable and effective means of earthing have been confirmed before any connections are made.
- **Temporary Electrical Installation for Structures:** Amusement Devices and Booths at Fairgrounds, Amusement Parks and Circuses.
- Electric Vehicle (EV) Charging Installation PME earthing facility shall not be used for supplying an EV charging point
- Onshore Units of Electrical Shore Connections for Inland Navigation Vessels ESQCR prohibits the connection of a PME earthing facility to any metalwork in a boat

Related topic questions

There are many aspects to exam success; probably the most significant is to read very carefully the question and "tease-out" the more noteworthy word or words that can be of assistance in determining a suitable answer.

Q7. Who is responsible for recommending the first Inspection and Testing after the Initial Verification?
- **a)** The inspector
- **b)** The designer(s)
- **c)** The constructor
- **d)** The client.

Answer: b) The designer(s)

Reasoning

This type of question could initially *wrong-foot* the candidate because it is the inspector who carries out the Inspection and Testing. Therefore the logical response would be to assume that the inspector has the responsibility to determine the first Periodic Inspection after the Initial Verification, however in the *Next Inspection Section* of the *EIC* it is the designer(s) who has (have) this responsibility (see *BS 7671 Appendix 6 page 462*).

Note: When completing a Periodic Inspection it is the inspector's responsibility to determine the next Periodic Inspection.

Q8. What action should the inspector take if the test results deviate from BS 7671 Regulations 120.3 and/or 133.5 when conducting an *Initial Verification*?

a) Inform the installation's constructor

b) Re-test the installation

c) The inspector should apply the requirements of BS 7671

d) Forward the test results to the designer for verification.

Answer: d) **Forward the test results to the designer for verification.**

Reasoning

An electrical installation must comply to the Standard, however there may be instances where the designer has specified certain requirements for a given installation, therefore the inspector can either compare his/her Initial Verification test results with the designer criteria or forward the test results to the designer for verification (see GN 3 *Relevant Criteria page 18*).

Q9. Which IP Code should be applied to barriers or enclosures of factory-built equipment during Initial Verification?

a) IP 4X or IPXXD

b) IP 2X or IPXXB

c) IPXXD & IPXXB

d) IP Codes are not applicable.

Answer: d) **IP Codes are not applicable.**

Reasoning

Confirmation of IP Codes for barriers and enclosures *does not apply* to factory-built equipment during Initial Verification; however this exception *does apply* to site constructed equipment but it is rarely necessary to confirm compliance.

(see GN3 paragraph 2.6.11 page 56).

Q10. When an inspector carries out the inspection of a consumer unit for basic protection, which IP Codes are applicable when verifying the barriers within the enclosure?

a) IP4X

b) IPXXD or IP4X

c) IPXXD

d) IPXXB or IP2X.

Answer: d) **IPXXB or IP2X.**

Reasoning

This type of question can be misleading in the respect that it refers to *basic protection*. But the question also refers to *barriers within the enclosure*, which are there to prevent contact with live parts: the concept of *basic protection*. Consequently, the answer must be d) IPXXB or IP2X (see BS 7671 *Regulations 412.2.2.3 page 70 and 416.2 & 416.2.1 page 74*)

Q11. Which level of <u>IP protection</u> is required for a <u>flush mounted,</u> low-voltage, <u>switch</u> mounted outside Zone 2 in a bathroom?
   a) IP2X
   b) IP4X
   c) IPXXD
   d) IP4X and IPXXD.

Answer a) IP2X.

Reasoning

This type of question could generate a problem because it refers to a bathroom; however the switch is <u>outside Zone 2</u>. Therefore it can be considered to be an <u>enclosure</u>; consequently BS 7671 *Regulation 416.2.1* will be applicable.

Q12. An inspector is required to confirm the connection of the <u>main equipotential bonding</u> conductor to the <u>incoming gas supply metallic pipework</u> conforms to the requirements of BS 7671. Therefore the inspector must confirm the conductor is connected:
   a) Within 600 mm of the meter outlet union or at the point of entry into the building
   b) Where practical, within 600 mm of the meter outlet union at the point of entry into the building
   c) Where practical, within 600 mm of the meter outlet union
   d) Where practical, within 600 mm of the meter outlet union or at the point of entry to the building if the meter is external, provided there is no branch pipework.

Answer: d) Where practical, within 600 mm of the meter outlet union or at the point of entry to the building if the meter is external, provided there is no branch pipework.

Reasoning

Initially, any of the answers could be correct, however if the final sentence of *Regulation 544.1.2* is reviewed, it clearly states: *where practical within 600 mm of the meter outlet union or at the point of entry to the building if the meter is external*; whereas the final words of the *Regulation's* second sentence confirms the connection of the equipotential bonding conductor must be made before any *branch pipework*.

Q13. **When conducting a visual inspection of a device used for <u>additional protection</u>, the inspector is required to <u>verify its effectiveness</u> and to ensure it <u>complies with</u> BS 7671. How will the inspector achieve this objective?**
   a) An RCD with a rated residual operating current not exceeding 30mA with a warning notice to test the RCD six monthly
   b) A 30mA RCD with a warning notice to test the RCD quarterly
   c) An RCD with a rated residual operating current exceeding 30mA with an operating time not exceeding 40ms at a residual current of 5I∆n and a warning notice to test the RCD six monthly
   d) An RCD with a rated residual operating current not exceeding 30mA with an operating time not exceeding 40ms at a residual current of 5I∆n and a warning notice to test the RCD six monthly.

Answer: **d) An RCD with a rated residual operating current not exceeding 30mA with an operating time not exceeding 40ms at a residual current of 5I∆n and a warning notice to test the RCD six monthly.**

Reasoning

<u>Additional protection</u> can be provided with either supplementary bonding or an RCD. However, the question refers to <u>a device</u>, therefore it is the <u>effectiveness</u> of the latter that can be achieved with a series of electrical tests. Whether the device complies with BS 7671 will depend upon the residual operating current of the device (30mA) and its disconnection time (40ms) at a given residual current (5I∆n). Finally, an appropriate warning notice must be posted by the inspector (see *BS7671 Regulation 415.1.1 page 73 and 643.8 page 235*)

Note: The disconnection time, for a residual test current of 5I∆n, is not recorded on the Schedule of Test Results.

Q14. **There is a requirement for <u>additional protection</u> for a <u>conduit wiring system</u> other than an RCD. Which of the following will meet this requirement?**
   a) Supplementary equipotential bonding
   b) Earthing
   c) ADS
   d) Double or reinforced installation.

Answer: **a) Supplementary equipotential bonding.**

Reasoning

Unfortunately *Supplementary equipotential bonding* is possibly the Cinderella of additional protection, probably because it is used for ADS fault protection (see *BS7671 Regulation 415.2 page 74*)

Q15. The Initial Verification of a domestic installation is dated 1997 with a <u>Periodic</u> <u>Inspection</u> <u>recorded in 2007</u>. A <u>ring final circuit</u> was added to the installation in 2014 <u>with solar panels installed in the following year</u>. *Electrical Installation Certificates (EICs)* were completed for each additional circuit and a Periodic Inspection was conducted in 2017. What <u>warning notices</u> should the inspector expect to see fixed in a prominent position at or near the origin of the installation in order to comply with the current version of BS 7671 Wiring Regulations?

   a) Notice of Periodic Inspection and Testing
   b) Notice of Periodic Inspection and Testing and RCD six monthly testing Notice
   c) Notice of Periodic Inspection and Testing, RCD six monthly testing Notice and warning notice of non-standard colours
   d) Notice of Periodic Inspection and Testing, RCD six monthly testing Notice, warning notice of non-standard colours, alternative supplies and isolation.

Answer: d) **Notice of Periodic Inspection and Testing, RCD six monthly testing Notice, warning notices of non-standard colours, alternative supplies and isolation.**

Reasoning

The inspector has a responsibility to comply with BS 7671 Wiring Regulations, therefore the appropriate *Notices and Warning Notices* must be fixed in a prominent position at or near the origin of the installation. These are:

- **Periodic Inspection and Testing Notices** generated on completion of *Initial Verification* and *Periodic Inspections*.
- **RCD protection** should be provided for socket outlets not exceeding 32A. However, there are no exceptions for dwellings, they are compulsory; therefore an **RCD warning notice** must be posted. Furthermore, the new conductor colour code became mandatory for all new installations post 2006, consequently a **non-standard warning notice** is also required.
- The output from Solar Panels is regarded as an alternative power source, therefore **alternative supply warning notices** must be affixed in the appropriate locations, as indicated in BS 7671.
- An **Isolation Warning Notice** is generally a standard notice supplied with new consumer units (CU), nevertheless the inspector must confirm it is firmly in place in the appropriate position on the CU.

(see *BS7671 Part 5 Regulations 514.10.1, 514.12.1, 514.12.2, 514.14.1 & 514.15 pages 132 -134*).

Q16. The <u>continuity measurement</u> of a 3-phase induction motor circuit was recorded as an <u>open circuit</u> by the inspector during <u>the Initial Verification</u> of <u>industrial</u> installation. What action should the inspector take?

a) Make good the defect

b) Carry out a fault-finding procedure to locate the problem

c) Contact the installation's designer for the installation criteria

d) Contact the constructor to rectify the problem. Provided the defect is made good, the circuit will be retested before the EIC is issued.

Answer: d) Contact the constructor to rectify the problem. Provided the defect is made good, the circuit will be retested before the EIC is issued.

Reasoning

Any defects or omissions revealed by the inspector during an Initial Verification procedure should be rectified, but <u>not</u> by the inspector. IET Guidance Note 3 states: *it is not the responsibility of the person or organisation carrying out the Inspection and Testing to make good defects or omissions*. Therefore, the inspector should inform the installation's constructor in order to make good the defect; thereafter retest the circuit and, provided the problem has been resolved, issue the EIC (see *GN3 paragraph 2.6.2. item (d) page 40*).

Q17. The inspector will discuss with the <u>client</u> the <u>extent and limitations</u> of a <u>Periodic Inspection</u> before conducting the Inspection and Testing. Which part of the installation will <u>not</u> form part of the discussion <u>unless specifically agreed</u> with the client?

a) The reasons for the limitation

b) The reasons for any operational limitations

c) The electrical equipment in accessible roof space

d) Cable in trunking and conduits.

Answer: d) Cable in trunking and conduits.

Reasoning

The inspector will <u>always discuss with the client the extent and limitations</u> of the Inspection and Testing of an installation, more so with commercial and industrial installations. However, cables concealed in trunking, conduit, fabric of the building and underfloor do not generally come under discussion <u>unless the client specifically requests an inspection</u> (see *Section D of the EICR Appendix 6 BS 7671 page 473*).

Q18. During a <u>Periodic Inspection</u> the inspector observed there was <u>no supplementary bonding</u> on a conduit wiring system. What <u>Observation Code</u> should the inspector record?

a) C1
b) C2
c) C3
d) FI.

Answer: b) C2.

Reasoning

The absence of <u>supplementary bonding</u> could promote a <u>potentially dangerous</u> condition rather than one which makes the installation unsafe (*danger present*) with a risk of injury; the former requires <u>urgent remedial action</u> whereas the latter requires <u>immediate remedial action.</u>

Q19. An additional socket outlet has been installed in an existing ring final circuit located in a kitchen. Which is the most suitable document that should have been issued?

a) *Electrical Installation Condition Report*
b) *Electrical Installation Certificate*
c) *Minor Electrical Works Certificate*
d) Part P Certificate.

Answer: c) *Minor Electrical Works Certificate.*

Reasoning

In previous editions of the *Building Regulations: Part P*, kitchens <u>were considered</u> to be special locations. However, in the current version of the *Building Control Regulations* the only <u>special locations are those given in BS 7671</u>; therefore, a *Minor Electrical Works Certificate (Minor Works)* must be issued (see *Part P of the Building Regulations and BS 7671 Appendix 6 page 466*).

Q20. In which special location would an inspector be expected to test an RCD not exceeding 500mA?

a) A location containing a bath or shower
b) Marinas and similar locations
c) Agricultural and horticultural premises
d) Construction and demolition site installation.

Answer: d) Construction and demolition site installation.

> **Reasoning**
>
> The only special location in BS 7671, in which an RCD not exceeding 500mA is to provide protection for circuits supplying one or more socket outlets with rated current exceeding 32A, is Construction and Demolition site Installation (see *Section 704 Regulation 704.411.3.3.1 page 257*).

SUMMARY

The candidate must be thoroughly conversant with *all* aspects of the model Inspection and Testing documents given in BS 7671and highlighted in this chapter.

The Initial Verification and Periodic Inspection documents issued by the self-certifying organisation, such as NAPIT, NICEIC or ELECSA, do not necessarily conform *exactly* to those given in BS 7671. Therefore, regardless of the skilled individual's competence to complete his/her self-certifying Inspection and Testing documentation, it is essential to be fully conversant with BS 7671 documents because they will or could be referred to in the written exam.

The Inspection and Testing of an installation is to provide, where reasonably practical, a safe electrical environment; whereas the correct completion of the related certificates and reports provides documentary evidence of the installation's condition or its "*state of health*" throughout its lifespan, in *order to detect any deterioration, failings or defects; thereby affording an electrically safe environment.*

The candidate must be familiar with the correct terminology associated with Inspection and Testing documentation and understand the difference between a *Certificate* and a *Report*. The former effectively confirms an installation is electrically safe and allows usage, whereas the latter is informative and depends upon the professional opinion of the inspector to determine whether the installation is in an electrically safe condition *to continue to be used safely.*

Consequently, the inspector is a *Skilled Person competent to do the work (of inspecting and Testing),* furthermore the signatories to *Electrical Installation Certificate (EIC):* designer(s) and constructor, are also regarded as *Skilled Persons competent to do the work.*

3
Electrical safety

INTRODUCTION

The testing of any electrical installation, regardless of whether it is a *dead* or a *live* test, can involve an element of risk. Therefore the inspector must be prepared to recognise possibly potential hazards and the level of risk.

It is the inspector's responsibility, and duty, to safeguard not only his or her own safety, but that of others, when carrying out a testing procedure. There are a number of statutory and non-statutory documents that clearly explain the means of applying electrical safety measures. Accordingly they should be consulted, especially where electrical safety is paramount.

The significant requirements for electrical safety are highlighted in *Regulations 12, 13 & 14* in the *Memorandum of Guidance on the Electricity at Work Regulations 1989*.

3.1 Regulation 12 – *means for cutting off the supply and for isolation*

Regulation 12 explains in detail the necessary steps that should to be taken to prevent dangerous incidents arising, for example:

- The correct identification of the circuit to be inspected or tested before any undertakings take place.
- The identification of isolation points, which should be clearly marked and freely accessible.
- Measures that should be taken to prevent unauthorised interference or improper operation of the circuit or equipment by the means of isolation safety locks.

3.2 Regulation 13 – *precaution for work on equipment made dead*

When an installation, or part of it, has been made *dead* to allow work to be carried out, *adequate precautions* must be taken to prevent any part of it from becoming *electrically charged* or *made live*.

Regulation 13 identifies the actions which should be taken to ensure a circuit is *dead* and the means to prevent any re-energisation. These are:

- Ensure the correct circuit has been isolated.
- Establish the circuit is *dead*, using an *approved voltage indicator*.
- The functional operation of the *approved voltage indicator* must be checked against a *proving unit* immediately *before* the circuit's isolation is proven and immediately *after* the inspection and testing procedure has taken place.
- The *approved voltage indicator* must comply to HSE (Health and Safety Executive) Guidance Notes GS 38 (4th Edition|) *Electrical test equipment for use on low voltage electrical systems*.
- Locking-off the isolation device. If this add the word is between the words this & not to read: If this is not possible then remove, and retain, fuses and/or links.

3.3 Regulation 14 – *work on or near live conductors*

Ideally all circuits and electrical equipment should be de-energised, locked-off and proven *dead* before any work takes place; however there are circumstances where this may not be possible. Therefore suitable precautions, which are proportionate to the level of risk, should be taken to prevent injury, for example the inspector should:

- be competent to work on live equipment safely
- have adequate information to carry out the work on live conductors
- have suitable tools, equipment and protective clothing (where applicable)
- erect suitable insulated barriers or screens, place warning signs, take effective control of the work area, apply the recommended isolation practice and ensure the test instruments comply with GS 38.

3.4 Isolation procedure

The *Memorandum of Guidance on the Electricity at Work Regulations 1989* does give the inspector the necessary directions to ensure, where reasonably practical, a means of achieving a safe working environment.

Subsequently a safe isolation procedure has been developed, using the prominent points identified in the appropriate *Regulations* but not necessarily in a chronological sequence, which may *not* be required for all circumstances, as follows:

- Consult with the client in terms of: documentation, relevant information, associated hazards and any possible risks.
- Inform individual(s) who may be affected by the inspecting and testing procedure.
- Identify the circuit to be inspected and/or tested.
- Isolate the installation or part of it, lock-off and retain the key(s). If this action is not possible apply other isolating techniques.
- Erect suitable barriers.
- Post warning notices indicating electrical work is taking place.
- Test the installation, or a final circuit, is isolated using an *approved voltage indicator* which must comply with GS 38 (test between line conductors, line and neutral and neutral and earth where appropriate). The term "line" is used rather than "live", because the installation or a final circuit should be *dead/de-energised*.
- The functional operation of the *approved voltage indicator* should be checked *before* and *after* the isolation procedure, against a *proving unit* or a known reliable voltage source.
- On completion of the inspection and testing procedure the system should be re-instated, which in effect is the reverse of the above process with safety at the forefront.

Note: It is advisable, when a circuit has been identified, to confirm it is *live* before the circuit is locked-off; then check again when the circuit has been isolated to confirm it is *dead*. This action will assist in confirming the correct circuit has been isolated.

3.5 Terminology: electrical *charge* and *live*

The terms "*charged*" and "*live*" are not exactly the same, but can have a similar effect: to give an electrical shock.

Charged refers to an item of equipment which has acquired an electrical *charge* by either capacitive, inductive or static means and has retained this *charge* even though the equipment may have been disconnected from the rest of the system. Whereas *live* refers to the connection of an installation or part of it, to a source of electrical energy during normal usage.

Although the term "*dead*" is not defined in the *Electricity at Work Regulations*, a system can be declared *dead* if it is neither *live* nor *charged*.

Related topic questions

Q1. A 3-phase motor is to be installed, and the inspector will be required to safely isolate the motor's final circuit to ensure the electrical safety of the motor installers. List the steps, in the correct sequence, that the inspector must carry out to confirm the motor's final circuit is safely isolated.

1. Test the *approved voltage indicator* before and after the isolation procedure, against a *proving unit*. 2. Identify the correct circuit, its point(s) of isolation, isolate, lock-off and retain the key(s). 3. Erect suitable barriers and post warning notices. 4. Confirm circuit is dead with an *approved voltage indicator*, which complies with the GS *38* standard.

    a) 3, 1, 2 & 4
    b) 2, 1, 3 & 4
    c) 2, 4, 1 & 3
    d) 1, 2, 3 & 4

Answer: c) 2, 4, 1 & 3

Reasoning

There are always risks, and electrical hazards, when installing electrical equipment. Ideally they should be eliminated or at least reduced to a minimum; therefore the correct isolation procedure must be adopted.

    The inspector should always be conscious of his/her responsibility as a duty holder. Although the inspector is *not* installing the motor he/she is in control of the isolation of the motor's final circuit.

Q2. An inspector needs to safely isolate a final circuit, but the circuit has been isolated and locked-off by another person. What action should the inspector take?
    a) Attach a warning notice: working on circuit.
    b) Attach a safety lock to the existing lock.
    c) Locate the person working on the circuit and retain his/her key.
    d) Double lock the circuit and prove it is isolated before working.

Answer: d) Double lock the circuit and prove it is isolated before working.

Reasoning

Ideally the other person working on the circuit should have left a name and contact number on his/her warning notice and arranged for a multi-locking off hasp to be fitted with both parties' safety locks. If this procedure cannot be accomplished, then double locking and confirming the circuit is isolated should be adopted.

3.6 HSE gs 38: electrical test equipment for use on low voltage electrical systems

HSE GS 38 is a non-statutory document; the objective of the document is to give electrical safety guidance to inspectors and other electrically competent persons who are involved in testing of electrical circuits, diagnosis and repair of equipment. Furthermore, where electrical testing does *not* require the equipment or part of an installation to be *live*, it should *be made dead and safely isolated*; additional safety precautions should also be taken when using test instruments, such as:

- having a good understanding of the instrument to be used
- checking the instrument conforms to the appropriate British Standard safety specifications
- checking that test leads including any probes or clips used are in good order, clean, and have no cracked or broken insulation.

The guidance given in GS 38 on the safety requirements for *Test Probes and Leads*, is as follows.

Test probes

- They will have finger guards to prevent inadvertent hand contact with the *live conductors* under test.
- They will be fully insulated leaving exposed metal tips *not exceeding* 4 mm, ideally 2 mm or less, or spring-loaded retractable screened probes are acceptable.
- They should have a suitable high breaking capacity (hbc) protective fuse, usually not exceeding 500mA or a current limiting resistor and a fuse, except when used to conduct an earth fault loop impedance test or, when testing an RCD, a 10A fuse may be used.

Leads

They are:

- adequately insulated
- coloured so that one lead can be easily distinguished from the other
- flexible and of sufficient capacity for the duty expected of them
- sheathed to protect against mechanical damage
- long enough for the purpose
- sealed into the body of the voltage detector.

Although HSE GS 38 *Electrical test equipment for use on low voltage electrical systems* is not directly involved in the testing *procedure*, it is extremely important that the inspector is familiar with the parameters required of test instruments; moreover the instrument used for proving a circuit is *dead* or *live* is an approved voltage indicator.

Related topic questions

Q3. Electrical testing inherently involves some degree of hazard; therefore the inspector has a duty of care to take adequate precautions to safeguard not only <u>his/her own safety</u> but that of others. Which <u>non-statutory document</u> provides <u>safety procedures</u> to achieve this objective?

a) Electricity at Work Regulations
b) BS 7671 Wiring Regulations
c) Health and Safety Executive Guidance Note GS 38
d) IET On-Site-Guide.

Answer: c) Health and Safety Executive Guidance Note GS 38

Reasoning

The key words to focus on in this type of question are those underlined: <u>his/her own safety</u> (the inspector's), <u>safety procedures</u> and <u>observed</u>, because they are used in the non-statutory document.

(*Guidance Note 3 Section 1: General Requirements paragraph 1.1 page 11.*)

The primary consideration when testing either an individual circuit or the entire installation should be the safe isolation, then checking to ensure the circuit/installation is dead with the appropriate instrument, which must meet the standard required by GS 38, before considering any further action.

Q4. An inspector is to carry out an <u>external earth fault loop impedance</u> (Ze) test. Before conducting the test the inspector should ensure the instrument's <u>test probes meet the requirements of GS38;</u> which statement confirms compliance?

a) Finger guards, exposed metal tips of 4 mm and a protective fuse
b) Finger guards, exposed metal tips not exceeding 2 mm and a protective fuse
c) Finger guards, exposed metal tips not exceeding 4 mm and a 500mA HRC protective fuse
d) Finger guards, exposed metal tips not exceeding 4 mm and a 10A protective fuse.

Answer: d) Finger guards, exposed metal tips not exceeding 4 mm and a 10A protective fuse.

Reasoning

The external earth fault loop impedance (Ze) is a live test which can be hazardous. It also carries an element of risk, therefore finger guards are essential to protect the inspector from inadvertently touching a live conductor. Moreover, the exposed metal tips of the probes should not exceed 4 mm and be protected with a 10A fuse.

SUMMARY

The City & Guilds Chief Examiner has stressed, in previous exam Feedback Reports, the need for candidates to read exam questions carefully and to relate their answers to the given question. This advice is still relevant with the advent of C&G 2391.

Clearly, from the data given in the Feedback Reports, candidates were not reading exam questions thoroughly or, on the other hand, there is the possibility that candidates had not fully understood the core element of a question.

Therefore the candidate should rationalise the question: What is being asked of the candidate? For example: does the question relate to:

- the safety of the inspector?
- the safety of others on-site?
- the safety of the general public or a combination of all three?

Or, if the question relates to GS 38:

- what does this document refer to?
- is the question requesting information on the instrument's probes or leads or both?

HSE booklet *Electricity at Work: Safe Working Practices* (HSG85) summarises working dead as follows:

1   Identify the circuit or equipment to be isolated.
2   Identify all isolation points.
3   Switch off supply (isolation).
4   Lock-off the isolate circuit or equipment and retain the keys.
5   Erect barriers and warning notices.
6   Use an *approved voltage indicator* to confirm the circuit or equipment is dead.
7   The *approved voltage indicator* must be tested before and after the isolation procedure, against a *proving unit* to confirm the test instrument is operational.
8   Take precautions against any adjacent live parts.

This basic isolation procedure must be supplemented with additional information as required by the question.

4
Installation testing

INTRODUCTION

Probably the most constructive advice which can be given to a candidate sitting the C&G 2391 theory exams, when doubting whether an answer to a question on testing is the *correct* or *expected* response, is to image actually carrying out the practical testing *in situ*; however the candidate must bear in mind that they are being examined on their testing abilities as an inspector, *not* as the installation's constructor or designer.

Therefore it is extremely important the candidate grasps this concept, simply because any defects and/or omissions that may be identified or revealed during the *Initial Verification*, **must not be rectified by the inspector**. The inspector should instead take the following action:

- Note the problems and report them.
- The "snags" must be made good, but *not* by the inspector.
- The installation must be *inspected and re-tested again* after the problems have been resolved and rectified.

Note: An *Electrical Installation Certificate* must *not* be issued until all defects and/or omissions have been made good by the constructor, and the installation meets BS 7671 standard.

4.1 Initial verification

The value of each measured test result must be recorded on a *Schedule(s) of Test Results*, and subsequently compared against the designer's criteria. One of the objectives of this exercise is, for example, to ratify the installation's earth fault loop impedance (Zs) values provided by the designer to ensure the final circuit's disconnection times are met.

The inspector must be vigilant when comparing the measured test results with those provided by the designer, to ensure they have not been affected by *parallel paths* created through exposed-conductive-parts and/or extraneous-conductive-parts, such as:

Electrical Inspection, Testing and Certification, 9781138488816.
© 2018 M. Drury. Published by Taylor & Francis. All rights reserved.

- steel conduit
- steel trunking
- MICC
- steel wire armoured (SWA).

Note: *Before any testing* takes place the inspector must carry out a *detailed inspection* of the installation (see BS 7671 *Regulation 642.1 page 230*).

4.2 Sequence of tests

The sequence of tests, coupled with the *Initial Verification,* should be followed in numerical order for safety reasons and recorded on the *Generic Schedule of Test Results** given in Appendix 6 of BS 7671.

The inspector must also be familiar with:

- the series of tests and their sequence
- the instrument used for each test
- the test instrument's correct name, *not* a trade name
- the instrument's range and its SI measured units for each test. For example: *Insulation Resistance Tester:* megaohms (MΩ), *Low Resistance Ohmmeter:* ohms (Ω), *RCD Tester:* milliseconds (ms)
- the test method used
- any special precautions necessary for a particular test
- whether the test is to conducted with either power ON or OFF (*Live* or *Dead*)
- safe isolation – *Electricity at Work Regulations 1989*
- test leads and probes – GS 38 (4th Edition) *Electrical Test Equipment for Use on Low Voltage Electrical Systems.*

Note: ***The candidate *must be fully conversant* with the inspection and testing documentation given in BS 7671 for exam purposes.

4.3 Test sequence

1 Continuity of protective conductors:
   i) **protective conductors**
   ii) **ring final circuit conductors**
2 Insulation resistance
3 (i) Protection by SELV, PELV or (ii) by electrical separation †
4 Insulation resistance of non-conducting floors and walls ††

5 Polarity
6 Earth electrode resistance
7 Protection by automatic disconnection of the supply †
8 Earth fault loop impedance
9 Additional protection
10 Prospective fault current
11 Phase sequence
12 Functional testing
13 Voltage drop. †

Note: † These tests are not recorded on the *Schedule of Inspection*.

†† This is a specialist test.

4.4 Instrument check

Before any testing takes place, the following checks should be carried out on the instrument used for the given test, they are as follows:

- Is the instrument's calibration *in-date*?
- Do the leads comply with GS 38 (regardless of whether the test is either a *dead* or *live* test, the leads and probes should be checked)?
- The resistance of the leads must be *nulled or auto-nulled* when conducting a continuity test. If the leads cannot be *nulled*, their resistance should be noted and subtracted from the final readings (with multifunctional digital test instruments, the term *zeroed* is used).
- Has the correct range been selected on a multifunctional instrument (if used) for the given test?
- Record the serial number of the multifunctional instrument or the dedicated instrument's serial number.

4.5 Safe isolation

Before any *dead testing* takes place, the installation or final circuit *must* be safely isolated; however the type of response made by a candidate will, of course, depend on the format of the question. If, for example, the scenario to a question indicates that there are other people on-site, then the following procedure is recommended:

- **Communication** – Initially confirm with the client or other parties who may be directly affected by the removal of electrical energy that it is safe or convenient to carry out isolation.
- **Isolation** – Switch off the installation, using main switch.
- **Extended Isolation** – Switch off all circuit breakers or remove fuses.
- **Security** – Lock-off the installation or circuit breakers (if possible).
- **Caution** – Display warning signs.
- **Prevention** – Erect barriers where necessary.
- **Confirmation** – Prove the installation or final circuit is dead with the aid of an *approved voltage indicator* by testing between line conductors, line and neutral, line and earth and neutral and earth to ensure. The terms "test lamp" or "voltage tester" are *not used*.
- The *approved voltage indicator* must be verified against a *proving unit* before any testing takes place, to check the instrument is functioning correctly, and after isolation has taken place, for safety reasons to confirm the *approved voltage indicator* has not been damaged and is in good working order.
- **Re-instatement** – On completion of the testing, confirm with the relevant parties whether it is safe to turn the supply on.
- **Re-instatement** – Remove locks, warning notices, barriers, re-instate supply and confirm supply is on.
- **Functional Test** – confirm the final circuit is functioning correctly after re-instatement of the circuit's supply.

4.6 TEST 1 i): Continuity of protective conductors

Type of test: Dead

Test instrument: low resistance ohmmeter

The objective of Test 1 is to establish electrical safety via the confirmation of the protective conductors' continuity, ensuring they are *secure* and *correctly connected* thereby verifying a *sound electrical* system.

The types of protective conductors which require verification are:

- earthing conductor
- main bonding conductor(s)
- supplementary equipotential bonding conductors
- circuit protective conductors (cpc).

It is the responsibility of the inspector to maintain a high level of electrical safety *before* and *during* the testing procedure; to ensure the installation is correctly isolated and to confirm whether it is safe to disconnect any protective conductor, in particular the earth, before testing begins.

4.7 Continuity of protective conductor: testing methods

There are two testing methods which are recognised as being suitable to confirm the continuity of a protective conductor.

Test method 1

This is generally accepted as the standard test for confirming the *continuity* and the *polarity* of the *line* and *circuit protective conductors* (R_1 + R_2) for radial final circuits.

The testing procedure is as follows:

- Select correct instrument.
- Carry out instrument check (see Section 4.4).
- Complete the safe isolation process (see Section 4.5).
- Take a *flying lead* from the distribution board's (DB) or the consumer unit's (CU) internal main earthing terminal (MET) to the outgoing (load) side of the protective device of the final circuit under test; this effectively shorts the circuit's line conductor (R_1) and its cpc (R_2). (Also see GN3 page 43).
- Test, at *every relevant point*, between the circuit's line and cpc conductors.
- The measured impedance taken at the furthest distance, from the DB or CU, will not only be the *highest*, it will also be the *value recorded* on the *Schedule of Test Results* as (R_1 + R_2).

Test method 2

This is frequently referred to as the *long lead* or the *wandering lead* test simply because a length of supplementary cable is used as part of the testing procedure.

It is also a *dead test*, and generally used to confirm the continuity of *earth and bonding conductors*, although it can be used to confirm the continuity of a final circuit's *circuit protective conductor (cpc)*. If this method is used to measure a final circuit's cpc, its ohmic value will be recorded on the *Schedule of Test Results* under the heading of R_2.

The testing procedure for the *earth and bonding conductors* is as follows:

- Select correct instrument.
- Carry out instrument check (see Section 4.4).
- Complete the safe isolation process (see Section 4.5).
- Remove the earth conductor from the installation's MET.
- Connect one end of the *wandering lead* to the disconnected earth conductor, with its opposite end connected to one of the instrument's leads.
- The instrument's other lead attaches to the *secured* earth or bonding conductor, for example the bonding conductor on the incoming water or gas pipes.
- On completion of the test preliminaries, the continuity of earth or bonding continuity can be tested.
- On completion of the continuity testing procedure the earth or bonding conductor must be re-connected to the MET.
- Re-instatement of the supply (see Section 4.5).

If this continuity testing technique for a bonding conductor cannot be achieved, possibly because the bonding connection is not accessible because the bonding clamps have been "built in", then the following method should be adopted:

The leads of the low resistance ohmmeter should be connected between any two points on extraneous-conductive metallic pipes; and provided the measured value is within the order of 0.05 ohm continuity will have been confirmed.

(see GN3 page 45)

Note: Disconnecting the earth and bonding conductors from the MET is an attempt to reduce or possibly eliminate parallel paths, which can produce misleading results (see Section 4.1).

Although the *ohmic values* of the *earth* and protective *bonding conductors* are not recorded on the *Electrical Installation Certificate*, the following must recorded on the document with a simple tick ($\sqrt{}$):

- The continuity and its secure connection has been verified.
- The type of material used in their construction.
- Their cross sectional area (csa) in millimetres squared (mm²).

Bonding Conductors

These are terminated at a DB's MET, to provide a common point for all extraneous-conductive-parts and to maintain them at the same potential.

The type of extraneous-conductive-parts that need to be *bonded-down* can include all or a number of the following:

i) metallic water pipes

ii) gas pipes

iii) oil pipes

iv) central heating and air conditioning systems

v) exposed metallic structural parts of the building.

Note: Provided there is an insulation section incorporated into metallic pipes, at the point of entry into the building, it is not necessary to connect them to protective equipotential bonding

(see *BS7671 Regulation 411.3.1.2 page 58*)

Supplementary equipotential bonding

This type of bonding connects all exposed-conductive-parts, such as metallic conduit, trunking and pipes in the installation, to the installation's MET. The testing process is reasonably straightforward:

- it is not necessary to disconnect any protective conductors
- the long lead is connected to the MET
- the other instrument's lead is connected to exposed-conductive-parts and the continuity is checked.

If metallic conduit, trunking or steel wired armour (swa) is used as a *protective conductor*, the following procedure should be employed:

i) the complete length of the enclosure must be inspected to confirm it is intact

ii) *Test Method 2* should be used to confirm continuity.

4.8 Test 1 ii): Continuity of ring final circuit conductors
Type of test: Dead
Test instrument: Low resistance ohmmeter

A *ring final circuit*, commonly referred to as "the ring", is probably *the* most common domestic power circuit; therefore it could possibly be a frequently asked exam question.

The objective of this test is to confirm:

- the continuity of the line, neutral and cpc conductors
- each associated pair of conductors form a *ring*
- there are no interconnections within the *ring*.

There are three specific stages which must be followed to achieve the objective of Test 2. The *Chief Examiner* for *City & Guilds* has recommended the use of *simple sketches*, which are illustrated below and in GN3 paragraph 2.6.6 pages 45-49.

Step 1

- Use an *approved voltage indicator*.
- Check the instrument against a *proving unit*.
- Isolate supply.
- Confirm circuit is *dead*.

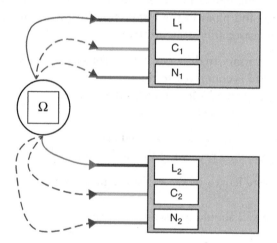

Figure 4.1

- Test the *approved voltage indicator* before and after confirming the circuit is dead with *a proving unit*.
- Disconnect all the *ring's* conductors.
- Use *low resistance ohmmeter*.
- Confirm continuity of each pair of *ring conductors* with an end-to-end test, that is L_1-L_2, N_1-N_2 and C_1-C_2. Each test values will be recorded as r_1, r_n and r_2 respectively on the *Schedule of Test Results* (see Figure 4.1).

Step 2

- Cross connect L_1 to N_2 and L_2 to N_1 (see Figure 4.2).
- Check the ohmic value at each socket outlet with the *low resistance ohm-meter* across the *line* and *neutral* terminals.

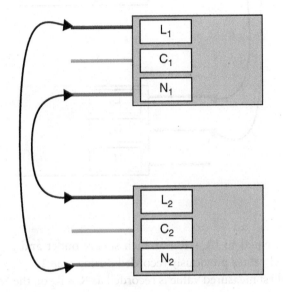

Figure 4.2

- The ohmic value at each socket outlet should be approximately the same.
- The value at each socket outlet should be

$$\frac{r_1 + r_n}{4}$$

where r_1 and r_n are the recorded values measured during Step 1.

Step 3

- Cross connect L_1 to C_2 and L_2 to C_1 (see Figure 4.3).
- Check the ohmic value at each socket outlet.with the *low resistance ohmmeter* across the *line* and *cpc* terminals.

- The ohmic value at each socket outlet should be approximately the same.
- The value at each socket outlet should be

$$\frac{r_1 + r_2}{4}$$

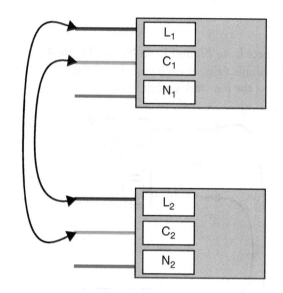

Figure 4.3

which is equal to $(R_1 + R_2)$ at each socket outlet and r_1 and r_2 are the measured values previously recorded during Step 1.

- The highest measured value is recorded as $R_1 + R_2$ on the *Schedule of Test Results*.
- On completion of the three stages, re-connect all the conductors.
- Safely re-connect power to the circuit.
- Carry out a functional and polarity check at each socket outlet.

4.9 Spurious results

The inspector may encounter a number of unusual or abnormal results when conducting Test 1: ii) *Continuity of Ring Final Circuit Conductors*, typically:

- A sudden increase in the ohmic value of a particular socket outlet during either *Step 2 or 3*: this is more than likely a spur, a visual inspection of the internal connections of the socket outlet will usually confirm this.
- There are *significant increases* in the ohmic values for a number of socket outlets and a similar decrease in the remaining socket outlets during either *Step 2 or 3*: the problem here is more than likely the incorrect cross connections of the relevant conductors, inasmuch:
  - *L_1 to N_1 and L_2 to N_2 or L_1 to C_1 and L_2 to C_2.*

Whereas the conductors should be:

- L_1 to N_2 and L_2 to N_1 or L_1 to C_2 and L_2 to C_1.

This type of problem normally occurs when cross connecting *single* core cables (see Figure 4.2).

- There is a *gradual increase* in the ohmic values for a number of socket outlets and a similar decrease in the remaining socket outlets during either *Step 2 or 3*: this is generally related to the characteristics of the copper conductors, which may not have a uniform cross-sectional area throughout its length and there is also a possibility that the ambient temperature may contribute to fluctuations in the measured values.
- Unexpected high ohmic values at possibly one or more socket outlet; this problem can be attributed to poor conductor terminations.

4.10 Test 2: Insulation resistance

Type of test: Dead

Test instrument: insulation resistance tester

The function or purpose of the *insulation resistance test* is to:

- determine any possible insulation break-down between conductors
- determine any possible ruptures, damage or deterioration in the conductor's insulation
- determine if any short circuits exist between conductors.

The instrument used for this test is an *insulation resistance tester* and it should have three dc voltage settings with an output current not exceeding 1mA. The voltage settings and application are as follows:

- 250 V dc for SELV and PELV circuits
- 500 V dc for all circuits where the installation's electrical supply is rated up to and including 500 V, but excluding extra-low voltage circuits
- 1,000 V dc for circuits where the supply voltage is rated at and above 500 V up to 1,000 V ($\geq$ 500 V and $\leq$ 1,000 V).

4.11 Testing preliminaries

Because Test 2 is a *dead test* the standard installation isolation process must be followed (see Section 4.5) and the standard instrument checks should be carried out (see Section 4.4) *before* any testing takes place.

Insulation resistance testing should take place during the installation's construction stage (1st Fix), mainly to confirm the integrity of the installed cables, more so with large installations; and re-tested as part of the *Initial Verification* procedure.

The testing process is as follows:

- Instrument check.
- Safe isolation of the installation, where an electrical supply is present.
- Ensure all protective conductors are terminated correctly, including the armouring (swa) and sheath of cables.
- Ensure any vulnerable equipment, electronic devices, lamps, central heating controls and RCDs are removed or isolated, to prevent damage to the equipment and to avoid any deceptive results.
- Main switch is OFF (open) and all circuit breakers are ON (closed) and/ or fuses are in place.
- All light switches are ON; where two-way and/or intermediate switching is part of the installation, the switches should be closed alternatively.
- Use an *Insulation Resistance Tester*.
- Select the appropriate voltage range. Although the test voltage for low-voltage installation is 500 V dc, it may be prudent to initially test at 250 V dc just in case any vulnerable or voltage sensitive equipment has been inadvertently missed and remains connected, thereby reducing the risk of any permanent damage.
- With all breakers closed, test between *live conductors* (Line and Neutral bus bars). Record the values on the *Schedule of Test Results*.
- Test between Line & Earth and Neutral & Earth Bus bars. Record the values on the *Schedule of Test Results*.

4.12 Test results

The *minimum* acceptable overall insulation resistance value is 1MΩ, however this value for a new installation should not be expected or acceptable. If a value of this magnitude or less is measured, then the following action should take place:

i) Open the individual circuit breakers one at a time when testing the *Line to Neutral* or *Line to Earth* conductors.
ii) When each circuit breaker is opened the installation is re-tested.
iii) When the resistive value returns to an acceptable level, on the re-test, the faulty circuit will have been located.

iv) If a value of less than 1MΩ is recorded, when measuring the installation's overall insulation resistance between its neutral and earth bus bars, then the inspector must endure the tedious task of disconnecting either the final circuit's neutral or cpc conductor and re-testing on each disconnection until the resistive value returns to an acceptable level.

Having located the final circuit causing the insulation resistance problem, the inspector must ensure all the conductors of the *sound* final circuits are re-terminated, in the correct numerical order, and conduct a re-test to confirm the problem has been resolved.

v) Where equipment and/or voltage sensitive devices are vulnerable to the test voltage and cannot be isolated or switched off, it is acceptable to connect together the line and neutral conductors and test the connected pair to the earth. It is essential that the incoming earth is connected to the installation main earthing terminal (MET). (see GN3 pages 49–53)

4.13 Test 3 i): Confirming SELV or PELV circuits by insulation testing

The inspector is advised to be fully conversant with BS 7671 *Regulation 414: Protective Measure: Extra-Low Voltage provided by SELV or PELV* before carrying-out Test 3. The objective of this test is to confirm the basic insulation integrity between either SELV or PELV circuits and adjacent low voltage circuits.

4.14 Testing procedure

The testing procedure is as follows:

- Select correct instrument.
- Instrument check (see Section 4.4).
- Isolate installation (see Section 4.5).
- Test the insulation between the line conductors of the SELV and/or PELV circuits and live conductors of adjacent low voltage circuits.
- Test voltage should be 250 V dc and the minimum acceptable insulation resistance is 0.5MΩ.
- Test the insulation between all SELV live parts and earth.
- Test voltage should be 250 V dc and the minimum acceptable insulation resistance is 0.5MΩ.
- If the SELV and PELV conductors are separated by basic insulation only, for example within a multicore cable with low voltage circuits, then the

test voltage shall be increased to 500 volts dc and the insulation resistance shall not be less than 1MΩ .

4.15 Test 3 ii): Electrical separation

Before carrying out the testing procedure the inspector should be fully conversant with *Regulation 413 Protective Measure: Electrical Separation* of BS 7671; also confirm the supply voltage is no greater than 500 volts.

There are three insulation tests which should to be carried-out for Test 3 (ii). They are:

- **Basic separation** – confirm the basic insulation between the primary and secondary live conductors of the isolating transformer is ≥ 1MΩ where the test voltage is 500 volts dc.
- **Basic insulation of the separated conductors** – confirm the insulation between electrical separated live conductors (secondary) and their corresponding exposed conductive parts is ≥ 1MΩ where the test voltage is 500 volts dc.
- **Basic insulation of any exposed-conductive-parts associated with separated conductors** – confirm the insulation between any exposed-conductive-parts associated with the secondary output circuits of the isolating transformer, and any protective conductor, other exposed conductive parts or earth is ≥ 1MΩ where the test voltage is 500 volts dc.

Before the above tests are carried out, the safety procedure below should be followed:

- Instrument check (see Section 4.4).
- Isolate installation (see Section 4.5).
- Confirm, with a *low resistance ohmmeter*, all exposed-conductive-parts of the separated circuits are bonded together
- Inspect all socket outlets to ensure the protective conductor contact is connected to the equipotential bonding conductor.
- Inspect all flexible cables to ensure that they contain a protective conductor for use as an equipotential bonding conductor (excluding feeds to Class II equipment).
- Test for Zs between live conductors, although this will only give an approximation of the prescribed live-cpc test. The recorded value(s) can be compared with the earth fault loop impedance Tables 41.2 and 41.3 in BS 7671 to confirm whether the measured Zs is acceptable to ensure safe operation of the protective device.

4.16 Test 4: Insulation resistance/impedance of floors and walls

Type of test: Dead

Test instrument: magneto-ohmmeter or battery powered insulation resistance tester

The magneto-ohmmeter or battery powered insulation resistance tester will provide a no-load voltage of approximately 500 V (or 1,000 V if the rated voltage of the installation exceeds 500 V) is used as a dc source.

This type of protection is regarded as a specialised form and is *not* recognised for general application, however where fault protection is provided by a non-conducting location, the following should be verified, prior to carrying out insulation testing:

- Exposed-conductive-parts should be inspected to confirm that no one can come into simultaneous contact with:
  i) two exposed-conductive-parts, or
  ii) an exposed-conductive-part and any extraneous-conductive-part.
- In a non-conducting location there must be no protective conductors.
- Any socket outlets installed in the location must not incorporate an earthing contact.

Expected results

The floors and walls are considered to be non-conducting where the measured resistances are at least 50 kΩ (where the system voltage to earth does not exceed 500 V).

If insulation has been applied to the extraneous-conductive-parts, within the location, during construction its insulation resistance must be tested. A test voltage of 2 kV ac rms is applied, with a *flash* insulation tester and any leakage current is measured, which should not be more than 1 mA. This test is applied on completion of the standard 500 V insulation resistance test.

4.17 Test 5: Polarity testing

Test instruments: (i) low resistance ohmmeter for dead tests; (ii) an Approved voltage indicator for live testing

A polarity test must be carried-out to ensure that the circuit's *line* conductor is connected in series with:

- the circuit's protective device, e.g. circuit breaker or fuse
- the single pole control, e.g. light switch
- every fuse within the circuit
- the centre contact of an Edison Screw type fitting, if applicable, except for either E14 or E27, which are not polarity sensitive
- the polarity of all socket outlets and similar accessories must also be confirmed.

Dead test

The *polarity* of circuit can be confirmed with the aid of either Test Methods 1 or Test 2, depending on the type of final circuit being tested.

If *Test Method 1* is used during Test 1then both the continuity and the polarity $(R_1 + R_2)$ of the line and cpc can be determine for radial circuits. Similarly, applying *Step 3* during Test 2, for ring final circuits, will also confirm both the continuity and the polarity $(R_1 + R_2)$; whereas *Test Method 2* (wandering or long lead method) can be used to determine just the polarity requirements of the line conductor (R_1).

A *Low Resistance Ohmmeter* must be used regardless of which Test method is used to determine the polarity of a final circuit.

Live test

For safety reasons, a live polarity check on a final circuit or an installation should be conducted on completion of the full Test sequence, normally as a functional test, and the instrument used is an approved voltage indicator.

4.18 Test 6: Protection by automatic disconnection of supply (ADS)

Fault protection (indirect contact) in a TN earthing system is provided by automatic disconnection of supply; and to ensure its effectiveness the following components need to be verified: the measured value of the earth fault loop impedance (Zs) meets the required standard; confirmation, by visual inspection, that the overcurrent devices used have suitable short-time or instantaneous tripping setting for circuit-breakers or current rating (In) and type for fuses electrical testing of installed RCDs to confirm they meet the required disconnection times

The effectiveness of ADS in a TT earthing system is verified in a similar manner as for a TN system.

4.19 Test 7: Earth electrode resistance

There are three methods which may be used to measure an earth electrode resistance. Each one is unique to a given application; they are classified as E1, E2 or E3.

- **E1 Application** – Generally accepted as the standard earth electrode test for generators and transformers.
  - **Test Instrument**: Earth Electrode Tester, usually the four-terminal type (fall of potential or three-terminal type are also available).
  - **Type of Test**: Dead.
- **E2 Application** – Depends upon the number of earth electrodes.
  - **One coil type instrument**: Used to test systems with just one earth electrode.
  - **Two coil type instrument**: Used to test systems with more than one electrode.
  - **Test Instrument**: Earth Electrode Tester, either a stakeless type or a probe type.
  - **Type of Test**: Dead.
- **E3 Application** – Generally associated with TT earthing systems where RCD protection is present.
  - **Test Instrument**: Earth Fault Loop Impedance Tester
- **Type of Test**: Live.

E1 testing procedure

Test instrument: four terminal earth electrode resistance tester

i) Isolate the supply.

ii) Disconnect the earthing conductor from the earth electrode, provided it is safe to do so.

iii) With the four terminal (P1, C1, P2 and C2) instrument, short terminals P1 and C1 and connect a lead to the earth electrode to be tested.

iv) Place the *Current* test probe, C2, a distance which is equal to ten times the length of electrode under test, e.g. 10 × 3 m = 30 m, from the earth electrode to be tested.

v) The *Potential* test probe, P2, is placed mid-point between C2 and the electrode under test.

vi) A minimum of three test readings should be taken, the first with the potential test probe P2 in its original position then a further two readings with P2 moved either side of P2's original position. The distance should be equal to 10% of the distance between the electrode under test and C2. In the example given in iv) the distance will be 3 metres.

vii) The average of the three readings is calculated and must not be greater than 21Ω.

viii) On completion of the test the earth must be re-connected.

Caution: If there is just one earth electrode this may leave the installation unprotected against earth faults, therefore complete isolation of the installation must be made.

E2 Testing procedure

Test Instrument: One test coil type

The testing method applied with the *one test coil type* of instrument is similar to that used for Method El, inasmuch as two temporary test spikes (electrodes) are used, but the earth electrode under test must *not* be disconnected.

Test Instrument: two test coil type

The testing method employed with the *two test coil type* of instrument is dependent upon the number of earthed electrodes within the installation; although the additional electrode(s) may not necessarily be the actual earth electrode(s) to be tested but extraneous-conductive-parts such as metallic services pipes or buried structural metalwork. The two coils of the instrument, one being a constant voltage variety and the other an induced current type, are clamped into place a small distance apart around the earthed electrode to be tested. The constant voltage coil induces a known voltage into a loop created by the earthed electrode and the general mass of earth; whereas the second coil measures the test current. It is not unusual, in practice, for the two coils to be combined in one clamp attachment.

E3 testing procedure

Test instrument: Earth fault loop impedance tester

i) Ensure it is safe to disconnect the earth conductor before testing takes place.

ii) Isolate the supply before disconnecting the earth conductor.

iii) Connect the leads of the instrument between the installation's incoming live conductor and the earth electrode.

iv) The impedance reading taken is treated as the electrode resistance.

v) On completion of the test the earth electrode must be re-connected. (see GN 3 pages 61–64)

Where an installation's earthing protection is provided by TT, the essential requirements to secure this type of protection are stipulated in *Regulations 411.5.1–411.5.4of BS 7671*.

A significant component in the TT protection system is the RCD. Where this device is used the following conditions must be fulfilled:

$$R_A \times I\Delta n \leq 50 \text{ volts}$$

Where:

R_A is the sum of the resistances of the earth electrode and the protective conductors connecting it to the exposed-conductive-parts [in ohms (Ω)].

$I\Delta n$ is the rated residual operating current of the RCD.

Where R_A is *not* known, it may be replaced in the above equation with Zs. Consequently the earth fault loop impedance (Zs) can be determined as follows, after the transposition of the original equation:

$$Zs = \frac{50}{I\Delta n}$$

where $I\Delta n = 30\text{mA}$

Note: $30\text{mA} = 30 \times 10^{-3} \text{ A} = 30 \times 0.001\text{A} = 0.03\text{A}$.

Therefore Zs = 50/0.03 = 1667Ω. (see BS7671 Regulation 411.5.3 page 64)

Currently there are no Zs Tables in BS 7671 for a TT earthing system, which highlights the maximum earth fault loop impedances, for circuits protected by circuit breakers or fuses nevertheless if *Zs exceeds 200Ω* it is considered to be *unstable*; that is, the disconnection times given in *Table 41.1* and *Regulation 411.3.2.4* of BS 7671 may not be achieved.

Note: Table 41.5, on page 64 of BS7671, only gives maximum Zs values for a TT earthing system associated with a given RCD

Furthermore, where an RCD is used for fault protection the circuit should also incorporate an overcurrent protective device (see *Chapter 43 of BS 7671*).

An overcurrent protective device may be used for fault protection, provided the final circuit's value of Zs is reliable, permanent, has a sufficiently low value and the following condition has been fulfilled:

$$Zs \times Ia \leq Uo \times Cmin$$

and
$$Zs = \frac{Uo \times Cmin}{Ia}$$

Where:

Ia can be considered to be the fault current in amperes (A), which will cause the protective device to operate within the time specified in *Table 41.1*.

Uo is the nominal a.c. rms or dc line voltage to earth in volts (V).

Cmin is the minimum voltage factor to take account of voltage variations depending on time and place, changing of transformer taps and other considerations.

4.20 Test 8: Earth fault loop impedance verification

It is essential, during the *Initial Verification* process, that the earth fault loop impedances are measured and that they are confirmed to be within the prescribed limits.

Because the earth fault loop impedances (Zs) are part of installation protective measure, it is imperative that the inspector is fully conversant with all the elements within an *earth fault current loop*; moreover, that they can identify the route of the fault current and the associated elements.

The starting point should be where the measured values are recorded on the line-earth loop as follows:

- the circuit protective conductor (R_2)
- the main earthing terminal (MET)
- the earthing conductor, which will be connected to either:
  - the metallic return path for TN-S systems, the SWA of the supplier's cable
  - the neutral of the supplier's cable for TN-C-S (PME) systems. The combination of the earth and neutral (PEN – Protective Earth Neutral) will create both a functional and a protective conductor

- an earthing rod for TT systems, using the main body of the earth as the earth return path.
- the path to the earthed neutral point (star point) of the transformer
- transformer winding (Tx)
- line conductor from the transformer winding to the supplier's protective device (Line)
- from the supplier's protective device to the consumer's protective device
- the line from consumer's protective device to the measured fault point (R_1).

Note: It is extremely important that the candidate fully understands the earth fault loop impedance pathway, can demonstrate this understanding and recognise either of the following earthing systems: TN-S, TN-C-S & TT.

4.21 Measurement of earth fault loop impedance (Zs)

Two methods may be employed to verify the total earth fault loop impedance for a circuit, they are:

Method 1

Test instrument: Earth fault loop impedance tester

This is a live test, therefore adequate safety precautions must be taken before any testing takes place; more so where there are other personnel on-site:

- Ensure all relevant personnel are informed of the live testing.
- If necessary erect safety barriers.
- Check the instrument's calibration is within date.
- Check the test instrument's leads conform to GS 38 and they are not damaged.
- Check the earth and bonding conductors are in place and secure.
- Test between line and cpc at every relevant point in the circuit.
- The reading taken at the circuit's extremity will not only be the highest value, it is also the value which is recorded on the *Schedule of Test Results*.
- Compare the measured values, take into consideration the correcting factors of 0.8 with the maximum tabulated values (see BS 7671 *Chapter 41*).
- On completion of the testing the relevant personnel should be informed and the barriers removed, if used.

Note: Although it is the responsibility of the inspector to carry out the earth fault loop impedance test, and to compare the test results with relevant design criteria or the tables given in BS 7671, it is *not* to fault find if the measured values do not meet the required standard.

Method 2

Test method: Use data from previous tests

The earth fault loop impedance (Zs) can also be determined using the measured values recorded from previous tests, with the aid of the following equation:

$$Zs = Ze + (R_1 + R_2)$$

Where $(R_1 + R_2)$ is the recorded value established during the continuity testing of ring or radial final circuits.

Whereas Ze is the external earth fault loop impedance.

Note: The inspector should be fully aware that parallel resistive paths may have been induced during the continuity testing of $(R_1 + R_2)$, which could cause a difference between the calculated and the measured values of Zs.

4.22 Measurement of external earth fault loop impedance (Ze)

Because the measurement of Ze is a live test, adequate safety precautions must be taken before any testing takes place; especially if there are others workers on-site. Subsequently, the inspector should carry out the following testing procedure:

- Ensure all relevant personnel are informed of the live testing.
- Erect safety barriers and warning notices.
- Check the instrument's calibration is within date.
- Check the instrument's leads conform to GS 38 and are not damaged.
- Switch off the supply at the distribution board or consumer unit.
- To remove the possibility of parallel paths, the earth conductor should be disconnected from the installation's earthed bonding conductors, providing it is safe to do so.
- Test between supplier's line and earth conductors at the origin of the installation.
- The reading taken is recorded on the *Electrical Installation Certificate* and the *Schedule of Test Results*, also on the *Electrical Installation Condition Report* when conducting a Periodic Inspection.

- On completion of the testing, the relevant personnel should be informed and the barriers removed.
- It is extremely important that the *earth* conductor is replaced before the main switch is closed.

Additional information

Large Installations: The *external earth fault loop impedance* (Ze) is expected to be measured at the origin an installation, hence the term "external"; however this term can cause confusion especially in large installations where there could be a significant number of sub-distribution boards (sub-DB).

The inspector needs to confirm the continuity of an "earth fault path" throughout the installation; therefore an *external earth fault loop impedance* measurement must be made at each sub-DB, and designated as Zdb rather than Ze to give the equation $Zs = Z_{db} + (R_1 + R_2)$.

Therefore the additional impedance, created by $(R_1 + R_2)$ of the supply cables to each sub-DB, should be recognised and taken into account.

Enquiry: The external earth fault loop impedance, Ze, may be determined by contacting the relevant Electricity Distribution Network Operators (DNO); however if this method is used the inspector will need to confirm the fault protection loop is continuous, which will, of course, require an earth fault loop impedance test!

4.23 Test 9: Prospective fault current (Ipf)

Test instrument: Earth fault loop impedance set to prospective fault current range

Type of Test: Live

The inspector is required to determine both components of the prospective fault current, which are:

- *Prospective Short-Circuit Fault Current (PSCC)*
- *Prospective Earth Fault Current (PEFC).*

The measurement of these fault current will involve live testing, therefore adequate safety precautions must be taken before any testing takes place; more so if there are other contractors or workers on-site. Accordingly, the inspector should carry out the following testing procedure:

- Ensure all relevant personnel are informed of the live testing.
- Erect safety barriers and warning notices.
- Check the instrument's calibration is within date.
- Check the instrument's leads conform to GS 38 and are not damaged.
- Switch off the supply at the distribution board or consumer unit.
- The earth conductor must *NOT be disconnected* from the installation's MET.
- Test between supplier's live and earth and live and neutral conductors at the origin of the installation, *PEFC and PSCC* respectively.

Note: On large installations the prospective fault currents should be measured at every relevant point in the installation.

- The highest prospective fault current measured will be recorded on the *Electrical Installation Certificate* and its *Schedule of Test Results*, also on the *Electrical Installation Condition Report* and its *Schedule of Test Results* when carrying out a Periodic Inspection.
- On completion of the testing the relevant personnel should be informed and the barriers removed and the system re-instated.

Note: The term "every relevant point" means every point where a protective device is required to operate under fault conditions, which includes the origin of the installation. (see GN3 page 68)

4.24 Reasons for measuring PSCC and PEFC

The reason for measuring the prospective fault currents is to ensure the short circuit breaking capacity (Icn) of a final circuit's protective device is not exceeded; therefore highest measured prospective fault current value must not exceed this value.

Note: If the breaking capacity of the protective device at the origin of the installation is not exceeded, and the protective devices downstream have a similar short-circuit rating, then no further checks are necessary simply because the magnitude of the prospective fault current will decrease as the circuit resistance increases.

4.25 Breaking capacities

There are two short-circuit capacity ratings:

- Icn – which is the maximum fault current a protective device can interrupt safely, although it may no longer be usable, and is the value marked on the face of the device in a small rectangle, for example 6,000 which is equal to 6,000 A or 6kA.
- Ics – the maximum fault current the device can interrupt safely without loss of performance.

The short circuit breaking capacities, and operating characteristics for both Icn and Ics are similar up to 6 kA but beyond this value there are a significant operating differences, notably:

- The protective device will operate safely for the value given on the device, although it must be replaced.
- provided the protective device's Ics value is not exceeded, the device can remain in use.

(See *BS EN 60898 & BS EN 61009 for definitions and IET Guidance Note 3. Table 2.8 pages 71 & 72*)

4.26 Test 10: Additional protection
Residual current devices (RCDs) testing
Test instrument: RCD tester
Type of test: Live

It is critical that the earth fault loop impedance (Zs), for the protected circuit, is determined *before* the RCD is tested; primarily for safety reasons, namely:

- The continuity of fault path is essential because without it the RCD will not function or could result in spurious action.
- If the measured earth fault loop impedance (Zs) does not meet BS 7671 standards there is a high possibility that the prescribed disconnection times for the protected circuit will not be achieved.

The testing should be made on the load side of the RCD, between the line conductor and the cpc of the protected circuit, with the load disconnected.

Preventative safety precaution should be taken when testing RCDs to ensure individuals or livestock do not come into contact with exposed-conductive-parts and extraneous-conductive-parts because the testing process can generate potentially dangerously high voltages especially when or if the earth fault loop impedance approaches its maximum acceptable limits.

Where an RCD is used for *additional protection* its rated residual operating current $(I_{\Delta n})$ should not exceed 30 mA. In the event of a basic and/or fault protection failure, the operating time of the RCD must not exceed 40 ms. Accordingly, the RCD must be tested at five times its rated residual operating current $(5I_{\Delta n})$ to confirm the device does actually disconnect within the 40 ms permitted time period.

Note: The maximum disconnection test time of 40 ms should not be exceeded; however if the protective conductor potential rises by less than 50 V there is a possibility that this maximum value may be exceeded; therefore the instrument's supplier should advise on compliance.

RCD Application

Additional protection (see BS 7671 regulation 411.3.3 (i) & (ii))

With the introduction of the *3rd Amendment (BS 7671:(2015))*, RCD protection *Regulations* have undergone significant modifications, inasmuch as there must be *additional protection* in every type of installation regardless of whether it is domestic, commercial or industrial, Regulation 411.3.3 has undergone further modification in the 18th Edition *Wiring Regulations* inasmuch all socket outlets, regardless of their installation position and not exceeding 32 A, must be protected by an RCD not exceeding 30mA.

A documented *Risk Assessment* is required where *additional protection* is not provided for socket outlets not exceeding 32 Amperes. The document must be attached to the *Electrical Installation Certificate* or the *Minor Electrical Installation Works Certificate* depending on which certificate is relevant.

The installation's designer or installer must justify the reason(s) why *additional protection RCDs* has not been fitted (see *BS 7671 Appendix 2* Item 11 page 360).

There is *no* exemption for domestic properties, *additional protection* must be provided.

RCD's integral test device

An integral test device is incorporated in every RCD and it is identified with either the word *Test* or the symbol *T* on a test button, which is used to verify whether the *mechanical parts* of the RCD are functioning correctly. Moreover, the mechanical test must take place when the RCD is energised.

The use of the integral test button is to check the *mechanical operation* of the RCD, it will *not* ratify any of the following:

- the continuity of the earthing conductor or the associated circuit protective conductors, or
- any earth electrode or other means of earthing, or
- any other part of the associated installation earthing, or
- the sensitivity of the device.

4.27 Test 11: Phase sequence

Type of Test: Live
Test Instruments Types:
Rotating Disc
Indicator Lamp

In a multiphase (3-phase) installation the phase sequence must be maintained *through-out* the installation, that is L1, L2, L3.

The inspector can confirm this configuration either visually or when conducting the polarity test, however it may be necessary substantiate the phase sequence using either of the test instrument types.

4.28 Test 12: Functional testing

The number of protective devices, which must be functionally tested, has gradually increased with the introduction of the current *Wiring Regulations*. Probably the most significant is the Arc Fault Detection Device (AFDD) which is recommended as a means of providing additional protection where there is a risk of fire. Those devices which have an integral test button should be used to test the mechanical function of the device, the result of the manual test <u>must</u> be recorded on the *Schedule of Test Results* (see GN3 paragraph 2.6.20 page 76 and BS7671 Regulations 421.1.7 & 532.6)

Although *Residual Current Monitors* (RCMs) and *Insulation Monitoring Devices* (IMDs) are not new to the 18th Edition their correct functioning must now be verified. Moreover, if a circuit is permanently monitored by these devices; and providing their operating function is proved to be correct, it is NOT necessary to measure the circuit's insulation resistance. (see BS7671 Regulation 651.2 page 237)

Note: Currently there are no requirements to record the correct functioning of either of these devices.

In addition to testing both the electrical and mechanical operation of an RCD, other electrical devices and assemblies, such as switchgear, controls, interlocks and light switches should be functionally tested or operated to confirm that they work, they are properly installed, mounted and adjusted. (see BS7671 Regulation 643.10 page 235)

4.29 Test 13: Verification of voltage drop

There has been a complete revision of the means of determining a circuit's voltage drop in the 18th Edition; the new Regulation simply states: the voltage drop shall be evaluated by *measurement* or by *calculation*. However the inspector is required to verify compliance with the Regulations in Section 525 inasmuch a final circuit's voltage drop does <u>not</u> exceed the values given Table 4Ab of Appendix 4 page 383 (Also see BS7671 Regulation 643.11 page 235)

Moreover the verification of a voltage drop is not normally required during initial verification; this is the responsibility of the installation's designer, although the design criteria may be used if required to determine a voltage drop

During a *periodic inspection and testing* the inspector is required to verify the voltage drop for each final circuit complies with BS7671 Section 525. To ensure compliance the inspector can either:

i)   measure the final circuit's impedance, apply Ohm's Law Equation and evaluate the voltage drop (see *Chapter 8* where $V_d = I \times R$) or

ii)   use the designer's criteria to calculate the voltage drop or

iii)   measure the voltage at the origin of the installation and at the terminals of the fixed current using equipment or at the socket outlet; the difference between the two values will give the voltage drop which should not exceed the values given in Table 4Ab of Appendix 4

Note: The designer's criteria will include diagrams or graphs clearly indicating, for example, the relationship between the following:

- maximum cable lengths
- conductor cross-sectional areas
- load currents
- percentage voltage drops
- conductor temperatures
- wiring systems.

Related topic question

Q1. State the course of action the inspector should take to determine the <u>voltage drop</u> of a radial power circuit during Initial Verification.

a) The voltage drop can be calculated using the designer's criteria.

b) The voltage drop can be calculated using the circuit's measured impedance.

c) The voltage drop is not normally required.

d) Measure the voltage drop.

Answer: c) The voltage drop is not normally required.

Reasoning

The inspector is <u>not normally</u> required to determine the voltage drop of a final circuit when conducting an <u>Initial Verification</u> , simply because this responsibility rests with the installation's designer.

4.30 Verification in medical locations

Initial verification

The introduction of *verification in medical locations* is relatively new and very much a specialist area; nevertheless the inspector will be expected to comply with the *Initial Verification procedure* as stipulated in BS 7671 Chapter 64, in addition to other tests associated with *Medical Locations Section 710*, and they are:

- The complete functional testing of IMDs (insulation monitoring devices) associated with the medical IT (Isolated Terra (Earth)) system equipment.
- Measurement of leakage current of the output circuit and of the enclosure of medical IT isolating transformers.
- Measurement and verification of the supplementary equipotential bonding resistance.

Periodic inspection and testing

The *Periodic Inspection and Testing* of *Medical Locations* will correspond with the requirements of Chapter 65, in addition to the following tests, at the recommended time intervals:

- Annually: Complete functional tests of the insulation monitoring devices (IMDs) associated with the medical IT system.
- Annually: Measurements to verify the resistance of the supplementary equipotential bonding.
- Every 3 years: Measurements of leakage current of the output circuit and of the enclosure of the medical IT transformers in no-load condition.

4.31 Temporary overvoltages due to high voltage systems

To confirm compliance with BS 7671 *Regulations 442.2.1 and 442.2.2*, the inspector will require appropriate information from the designer of the installation, and other parties, to enable him or her to verify compliance, so far as is reasonably practicable. Accordingly, the inspector will need to check the following:

- the high voltage and low voltage earthing arrangements of the substation have been correctly installed
- the earthing resistances to earth meet the designer's requirements
- confirm the high voltage earthing and low voltage arrangements are interconnected
- if separated, and where appropriate, meet the designer's requirements
- all earthing systems, and any additional earthing connections on the LV side of the installation, are correctly installed and their resistances to earth meets the designer's criteria
- ensure safety is not compromised
- the current rating of the protective devices and their settings meet the designer's requirements.

4.32 Verification of protection against overvoltages of atmospheric origin or due to switching

The inspector should be aware of his or her responsibility, during the *Initial Verification* process, inasmuch as all the installation's electrical equipment has been inspected to ensure it has been selected and erected (installed) in accordance with *BS 7671 Regulations* and supported, where necessary, with the equipment's manufacturer's instructions and product standards (see Regulation 642.2 (ii)).

Consequently, the inspector will be required, as part of the investigative procedure, to check whether the equipment can withstand the impulse voltages (Uw) given in BS 7671 *Table 443.2*, regardless of whether surge protective devices (SPDs) have been specified by the installation's electrical designer.

4.33 Selection and erection of surge protective devices

The requirement details and applications for fitting SPDs, within an installation, are given in *Section 534 of BS 7671*, whereby the application of a surge protective device, whether it is a Type 1, 2 or 3 will depend upon the protection required within a lightning protection zone (LPZ).

SPDs are required, possibly in collaboration with other devices, to limit:

- transient overvoltages of atmospheric origin, which can be transmitted via the supply distribution system
- overvoltages generated during switching operations
- transient overvoltages caused by direct lightning strikes (strokes)
- lightning strikes (strokes) in the vicinity of buildings protected by a lightning protection system.

Note: The term *stroke(s)*, which may be defined as an unexpected occurrence, is used throughout Section 534 of BS 7671 as opposed to the more common or colloquial term *strike* when referring to lightning.

SPD Types and application

Type 1 SPDs are fitted near the origin of an installation to prevent dangerous sparking/arcing, which can lead to fires or electric shocks.

Note: Type 1s are often referred to as equipotential bonding SPDs; however lightning protection systems, which rely on this type of protective SPD, do not provide effective protection against failure of sensitive electrical and electronic systems.

Type 2 SPDs may also be fitted at the origin of an installation, depending on the voltage stress levels, but generally used close to the protected equipment; more than likely at sub-distribution boards or consumer units as a means of protection against switching transients, which could be generated within a building.

Type 3 SPDs are generally incorporated in BS 1363 socket outlets as a means of protecting individual sensitive or critical electrical equipment, however the inspector should be aware, when conducting the insulation resistance test, that these devices are test voltage sensitive, which should be noted on the relevant diagram and documentation (see Regulation 514.9.1).

If the SPD cannot be disconnected during the testing process, the test voltage should be reduced to 250 volts dc with a minimum acceptable insulation resistance value of 1MΩ.

4.34 Verification of measures against electromagnetic disturbance

The inspector should give due consideration to the positioning of equipment which could be a source of electromagnetic disturbances, for example:

- switching devices for inductive loads
- electric motors
- fluorescent lighting
- welding machines.

Therefore the inspector should be familiar with *Section 444 pages 107–120 of BS 7671* in the respect: the means of alleviating electromagnetic interference (EMI) and achieving electromagnetic compatibility (EMC) via a high standard of design, installation of cables and equipment.

The routing of cables, their distance from other cables and the provision of equipotential bonding can alleviate EMI:

- Information technology (IT) and telecommunication cables should not be installed within 130 mm of discharge, neon and mercury vapour lamps.
- Segregate IT telecommunication cables from low voltage (LV) cables as follows:
  i) 200 mm in free space or no containment
  ii) 150 mm on perforated (tray) open containment
  iii) no segregation required in solid metallic containment.

(see Table A444.1 page 119)

Note: Table A444.2 specifies: the minimum separation between power and signal cable; and the minimum separation between signal cables and the related current carry capacity of the power cables

Periodic inspection: Testing

Although the testing *procedures* for a *Periodic testing* are similar to those of *Initial Verification testing*, the testing *sequence* does not necessarily have to be identical simply because the installation is already operational. Furthermore, provided all previously tested results are readily available and comprehensively documented the equivalent range and level expected for an *Initial Verification* may not be necessary.

Consequently, the fundamental objectives of a *Periodic Inspection* are:

- to ensure the installation is in a satisfactory condition
- to ensure the installation is safe to use
- to confirm the disconnection times, given in *Chapter 41of BS 7671*, are met
- to highlight any defects or deterioration.

Related topic question

Q2. The <u>testing sequence</u> for a *Periodic Inspection* and *Initial Verification* can differ, however under what circumstance <u>must</u> the test sequence for the two testing methods follow those given in BS 7671?
   a) The installation has not been previously tested
   b) The installation has been previously tested
   c) There are comprehensive test records
   d) The inspector's experience and knowledge.

Answer: a) The installation has not been previously tested.

Reasoning

To allow an electrical installation to be operational without it being initially tested would place the installation's owner in a precarious legal position, inasmuch as the owner would be contravening statutory regulations, that is: *Electricity at Work Regulations 1989*.

   Consequently, the owner should have the original *Electrical Installation Certificate* with related *Schedules* and, where applicable, *Minor Electrical Installation Works Certificate(s)*. There could also be *Electrical Installation Condition Reports* and/or its predecessors with the appropriate *Schedules of Inspection and Testing*.

Therefore, provided these documents are available it may not be necessary to carry out the full range of tests, for example a ring final circuit's Steps 1 & 2. Accordingly, the inspector can make a professional judgement, based on his or her experience and knowledge, on what electrical tests are appropriate for a particular installation (see GN3 Section 3.10 *Periodic Testing* page 98).

SUMMARY

The purpose of Periodic Inspection and Testing is to determine whether an installation is in a satisfactory condition to allow it to continue to be used in a safe way. This extremely important decision is based on knowledge and experience of the inspector, therefore the competency of the skilled person carrying out Periodic Inspection and Testing must rest on a number of factors, for example:

- electrical education and training
- practical skills
- knowledge: the ability to disseminate data in the context of the ongoing safety of the installation
- visual inspection skills, to determine the age of installation components and recognise signs of any deterioration
- good testing skills and experience of older installations.

Consequently, the abilities of a candidate will be fully tested, regardless of whether the individual is sitting either *C&G 2391-50 or 51*or both *C&G 2391-52* Inspection and Testing exams.

Therefore candidates must be fully prepared before they sit either or both theory exams. As part of this preparation the candidate is advised to read, and absorb, the information given in IET's *Guidance Note 3: Inspection & Testing*. Apart from providing valuable information on Inspection and Testing techniques and procedures, it is also a significant source of exam questions.

In addition to the *Guidance Notes 3*, the following Chapters in BS 7671 are compulsory reading, namely

- Part 6 Inspection & Testing
  - i) Chapter 64 *Initial Verification*
  - ii) Chapter 65 *Periodic Inspection & Testing*
  - iii) Appendix 6 *Model Forms for Certification and Reporting.*

Nevertheless, there are other Parts and Chapters in BS 7671 which should not be neglected, for example:

- Part 3 Chapter 31 Sections 313, 312 and 313
- Part 4 Chapters 41, 43 & 44 Sections 443, 444 & 445
- Part 5
- Special Locations.

During the testing process *safety* is paramount, not only for the inspector but all other personnel who could be affected by the testing procedure. Therefore, high on the priority safety list is the safe isolation of an installation or final circuits.

Probably the most simplistic approach to safe isolation is:

1 Identify the circuit to be isolated.
2 Identify any associated isolation points.
3 Check the circuit is live with an approved voltage indicator.
4 Switch-off the circuit's supply, lock-off the circuit with a safety lock, attach warning signs and erect safety barriers, if required.
5 Test the circuit to confirm it is dead using the approved voltage indicator.
   - The approved voltage indicator must comply with GS 38 and not be damaged.
   - Check the approved voltage indicator before the isolation process begins and after completing the isolation process, this action is necessary to confirm the test instrument is functioning correctly.

The candidate must also be familiar with the type of instrument used for each Test; equally its correct name, range and measured units, and they are:

Table 4.1 Instruments and their usage

Instrument	Application
Low resistance Ohmmeter	Continuity
4 volts to 24 volts s/c current 200 mA	Polarity
Resolution of 0.05 or less	
Measurement Unit: ohms (Ω)	
Scale 0.2–2.0Ω	
Insulation Resistance Tester	Insulation Resistance
Measurement Unit: megaohms (MΩ)	
SELV or PELV Test Voltage: 250 volts dc	
Minimum Insulation Resistance: 0.5 MΩ	
Supply $\leq$ 500 volts ac Test Voltage: 500 volts	
Minimum Insulation Resistance: 1.0MΩ	
Supply > 500 volts ac Test Voltage 1,000 volts	
Minimum Insulation Resistance: 1.0MΩ	

Instrument	Application
Earth Fault Loop Impedance Tester	Earth Fault Loop Impedance, Zs
Measurement Unit: ohms (Ω) RCDs could trip unless the instrument has a _D-lok_ or some other inbuilt protective device. This instrument generally has a dual function, whereby prospective fault currents can be measured. Measurement Unit: kilo amperes (kA)	External Earth Fault Loop Impedance, Ze
Prospective Short Circuit Current P-N (PSCC) Live test	Prospective Earth Fault Current (Ia) (PEFC)
Units – kA	P-E for a two lead instrument; P-N-E for a three lead instrument
Earth Electrode Resistance Tester	Earth Electrode
Use a four lead tester C_1, P_1, P_2 & C_2 P = potential, C = current Distance between the electrode under test & C_2 electrodes is equal to 10 × length of earth electrode under test. Dead test Measurement Unit: ohms (Ω)	
RCD Tester	RCD
Measurement Unit: milliseconds (ms) Time/current performance criteria for RCDs to BS EN 61008 & 61009.	

Table 4.2 RCDs – Time and current criteria for BS EN 61008 & 61009

Current (mA)	Trip Time (ms)	RCD Current Rating (mA)			
		30	100	300	500
½IΔn	greater than or equal to 2000 ($\geq$ 2sec)	✓	✓	✓	✓
IΔn	less than or equal to 300	✓	✓	✓	✓
5 IΔn	less than or equal to 40	✓	n/a	n/a	n/a
Where n/a = not applicable IΔn is the current rating of the RCD					

Further related topic questions

Q3. When measuring the <u>external earth fault loop impedance</u> (Ze) what should the inspector do with the main earthing conductor and the reason why?

a) Leave connected to confirm a reliable prospective fault path
b) Leave connected to take into consideration any parallel earth paths
c) Disconnect to reduce or eliminate parallel earth paths
d) Disconnect to remove the load capacity of the installation.

Answer: c) Disconnect to reduce or eliminate parallel earth paths.

Reasoning

Parallel earth return paths, which can be attributed to extraneous conductive or other metallic parts within the installation, can give rise to misleading test results and cast doubt on the fault path reliability (see *GN3 paragraph 2.6.5 page 42*).

Q4. When measuring the <u>prospective earth fault current</u> state what the inspector should do with the main earthing conductor and the reason why.

a) Leave connected to confirm the earth fault loop impedance
b) Disconnect to eliminate parallel earth paths
c) Leave connected to include any parallel earth paths
d) Leave connected to determine the maximum prospective fault current to earth.

Answer: d) Leave connected to determine the maximum prospective fault current to earth

Reasoning

A reasonably straightforward response, simply because when the installation is energised, and if a fault current is generated, the earth conductor <u>will be connected</u>; therefore the inspector needs to know what the *actual* prospective currents are to ensure the <u>final circuit's protective device can operate safely under fault conditions.</u> Any parallel paths must be included.

Q5. The <u>prospective fault current</u> (I_{pf}) should be measured at every relevant point in an installation. Under what circumstances will this not be necessary?

a) The prospective fault current is equal to 10kA
b) The prospective fault current is equal to or less than 6kA
c) Icn is equal to Ics through-out the installation
d) All the protective devices throughout the installation have the same Icn value as the protective device at the origin.

Answer: d) All the protective devices throughout the installation have the same Icn value as the protective device at the origin.

Reasoning

Provided the protective device's Icn is <u>not exceeded</u> at the <u>origin of the installation</u>, and the Icn value of each protective device is identical through-out the system, it will be a pointless exercise to measure the prospective fault current at every relevant point because the impedance down-stream from the origin will increase therefore the prospective current will reduce proportionally (see *GN3 Section 2 Paragraph 2.6.17 page 68*).

Q6. The continuity of the line, neutral and protective conductors of a ring final circuit were verified at the installation's distribution board. The ohmic values of r_1, r_2 & r_n were recorded on a schedule of test results, which were used to calculate R_1 & R_n and R_1 & R_2; however when carrying out Steps 2 and 3 of the continuity of the ring final circuit testing procedure, one particular outlet socket, when tested at its tubes, gave open circuit results. What conclusion(s) can the inspector make?

a) There was an open circuit

b) A poor ohmic contact

c) Transposition of the cps and neutral conductors

d) Transposition of the cps and line conductors.

Answer: c) Transposition of the cps and neutral conductors.

Reasoning

If the ring final circuit had an open circuit it would have been detected when carrying-out Step 1 of the testing procedure (no continuity); similarly if poor ohmic contact existed this would have also been detected during Step 1 (very high ohmic readings).

If the cps and line conductors had been transposed at the socket outlet, R_1 & R_2 would have been confirmed regardless of the transposition; however with neutral and circuit protective conductors transposed, the inspector would effectively be testing between R_1 & R_2 and *not* R_1 & R_n.

A simple sketch of a socket outlet's connections will illustrate the points raised in the question's reasoning.

Q7. When carrying out a sequence of tests, what test would the inspector consider necessary to safely isolate the complete electrical installation?

a) Earth Fault Loop Impedance

b) External Earth Fault Loop Impedance

c) Functional Testing of RCDs

d) Insulation Resistance of an individual circuit.

Answer: b) External Earth Fault Loop Impedance.

Reasoning

Although the External Earth Fault Loop Impedance is a live test, it does require the disconnection of the main earth conductor to eliminate parallel earth paths, which removes the safety aspect of the installation, both basic and fault protection; therefore the installation must be totally isolated during this particular test.

The inspector *must* replace the main earth conductor *before* switching on the supply (see GN3 *pages 65 & 67.*

Q8. What other test, other than the answer given in Q7, would the inspector consider necessary to safely isolate the complete installation?

a) Earth Fault Loop Impedance

b) Continuity of a circuit's cpc

c) Continuity of the main protective bonding

d) Continuity of supplementary bonding of a metallic conduit wiring system.

Answer: c) Continuity of the main protective bonding.

Reasoning

The reasoning for the answer given for Q8 is similar to that given for Q7 in the respect, if a bonding conductor is disconnected, as required by Test Method 2, any fault current induced in extraneous metalwork would leave the individual, who may be in contact with the metalwork, unprotected (see *GN3 pages 44 & 45*).

Q9. What is the minimum short circuit test current and voltage range for a low resistance ohmmeter?

a) 200A at 24 volts dc

b) 200A at 24 volts rms

c) 200mA at 4 volts to 24 volts rms

d) 200 mA 4 volts to 24 volts dc or ac

Answer: d) 200 mA 4 volts to 24 volts dc or ac.

Reasoning

See GN3 page 109 and Table 4.1 Instruments and their Usage.

Q10. When testing a non-time delayed 300mA RCD, what is the maximum disconnection time the inspector can accept for a BS EN 61009 RCBO?

a) 30ms

b) 40ms

c) 150ms

d) 300ms

Answer: d) 300ms.

Reasoning

See GN3 Table 2.9 page 75 and Table 4.2 RCDs Time and Current Criteria.

5
Initial verification

INTRODUCTION

The *Initial Verification* of an installation is to be carried out *before* it is put in to service, but what does *Initial Verification* actually mean? What does it relate to?

It is primarily the *first* inspection and testing of a new installation, or an addition or alteration to an existing installation to confirm it meets the standard required by *BS 7671 Requirements for Electrical Installation.*

To achieve this objective it will be necessary to conduct a systematic inspection and testing:

- During the construction of the installation (1st Fix).
- On completion of the installation (2nd Fix).

During the inspection and testing process, *precautions* should be taken to:

- avoid causing danger to persons and livestock
- avoid causing damage to property and installed electrical equipment.

5.1 The purpose of the initial verification

The *reason* or *purpose* of *Initial Verification* is to confirm, through the inspecting and testing procedure, that the installation complies with requirements of BS 7671 in terms of:

- *its design*
- *construction.*

5.2 Departures from BS 7671

If there is any departure from *Regulation 120.3* by the constructor, whether *intended* or otherwise, the designer must be informed, who will give the departure due considerations. If the designer sanctions the departure his or her decision should be noted on the *Electrical Installation Certificate*, however the safety of the installation must not be compromised and shall not *be less than that obtained by compliance with the Regulations.*

If, on the other hand, there are any departures from *Regulation 120.3* stipulated by the *designer*, the inspector should either ask for the design criteria or forward the test results to the designer for verification with the intended design. In the *absence of such data the inspector should apply the standard set out in BS 7671.*

Related topic questions

Q1. The purpose of Initial Verification is to confirm the:
   a) Installation complies with the requirements of BS 7671
   b) Installation has been safely installed and not deteriorated
   c) Installation has been designed correctly
   d) Design and construction of the installation meets the Standard and is safe to place into service.

Answer: d) design and construction of the installation meets the Standard and is safe to place into service (see *GN3 Section 2 Initial Verification paragraph 2.1 page 17*).

Q2. How does the inspector confirm an _installation complies with the design and construction requirements of BS 7671_?
   a) It is inspected and tested on completion of the installation
   b) It is tested on completion of the installation
   c) It is tested during construction
   d) It is inspected and tested during construction and on completion of the installation.

Answer: d) it is inspected and tested during construction and on completion of the installation (see *BS 7671 Chapter 6 Regulation 641.1 page 230*).

Q3. Give the _reason_(s) why an inspector should inspect and test an installation _during_ its construction period.
   a) The requirements of the Regulation have been met
   b) To confirm design cables
   c) The cables meet the designer's criteria throughout the length of the circuits
   d) To confirm the installer/constructor has installed the cables.

Answer: a) The requirements of the Regulation have been met (see *BS7671 Chapter 6 Regulation 641.1 page 230 & GN3 Section 2 Initial Verification paragraph 2.1 page 17*).

Reasoning

There is a practical element to each question; something an inspector would normally carry out as part of the Inspection and Testing procedure; nevertheless the candidate (inspector) must know <u>the reason(s)</u> why the Inspection and Testing procedure is carried out <u>during the installation's construction</u>.

5.3 Foremost inspection

The inspection process will normally take place with all, or that part of the installation which is to be inspected, disconnected from the supply. Subsequently, *before* any *testing* takes place, the *installed electrical equipment* must be *inspected* first to ensure:

- it complies with British and Harmonised standards (see *Regulation 511*)
- the equipment has been selected and erected to BS 7671 Standards (manufacturer's instructions may be taken into account)
- there is no visible damage or defect which could impair safety.

Although the inspection of the installed equipment is foremost, the overriding factor is electrical safety; therefore the installed equipment must be *safely isolated* before the inspection process begins, primarily to protect the inspector.

5.4 Inspection of installed equipment

Probably one of the major stumbling blocks a candidate may encounter is the tendency to overlook the fact that it is their competency as an inspector which is being verified. Accordingly, the candidate must *"don the inspector's provable cap"* whilst sitting the theory exam, which may require a degree of lateral thinking. In other words, what would the inspector *do* on-site?

Accordingly, if the candidate is questioned on the inspection of a piece of equipment, he or she should try to *visualise* the item as if they were on-site. For example, if a question relates to inspection of a distribution board (DB), what would the inspector be looking for?

Initially it will necessitate the inspector using one or more of his/her senses, that is: *sight, touch, hearing or possibly smell.* Then the logical procedure would be to check the DB for the following:

- is it secured?
- has it been suitable selected to meet the demands of the installation?
- is it suitably erected?

Having completed these preliminary checks, the next stage is to carry out a visual inspection of the enclosure:

- **Warning Notices** – are there any in place? They will be required, for example, where the voltage behind the enclosure is greater than 230 volts to earth or RCDs protection is present or where there is an alternative supply.
- **IP Code** – does the enclosure comply? Does the top horizontal surface comply with IP4X or IPXXD? Are the remaining parts of the enclosure compliant with IP2X or IPXXB?
- **Identification** – are the final circuits clearly identified?
- **SWA Cable Glands** – if they are present, are they secure and correctly fitted?

Once again, the candidate is strongly advised to read an examination question very carefully before making a response; do not automatically assume, when answering a question on Initial Verification, that the installation is connected to the public supply. If the question clearly states the supply has not been installed then the safe isolation procedure, as described in Chapter 3, will not be necessary.

Where an electrical supply is present at the distribution board, a safety isolation procedure will be necessary and care must be taken before removing the enclosure.

Having removed the enclosure what will the inspector be expected to inspect?

- **Cable Identification** – are the line conductors correctly identified with either a letter and numbers e.g. L1.1, L2.1 & L3.1 or brown sleeving (as necessary)?
- **Cable sequence** – are the line, neutral and cpc in the correct corresponding numerical sequence? Are any of the cables/conductors damaged?
- **Cable CSA** – are the conductor's csa compatible with the protective device? Are the conductors and the protective devices secured? Do they meet the designer's criteria?
- **Earthing** – does the cross sectional areas (csa) of the earth and cpcs meet the prescribed standard? If steel wire armoured (swa) cables are fitted, are they suitably earthed? Is there a "banjo" fitted to each cable? Are the swa glands suitable secured?
- **Supply Conductors** – are the supply conductors in the correct sequence L1, L2 & L3 (left to right)? Are they correctly coloured coded?

5.5 Inspector's responsibilities

The duty or *responsibility* of an inspector during the *Initial Verification* or the *Periodic Inspection* process is to:

1   Ensure that no person or livestock is subjected to any dangerous situation and no damage is caused to property.
2   Ensure that the results of the Inspection and Test procedure are compared with the designer's criteria and BS 7671.
3   Confirm compliance or non-compliance with BS 7671.
4   Review the installation and report on its condition.

In the event of a dangerous condition being revealed by the inspector, *during the Periodic Inspection* process, a written report should be given to the person ordering the work, which will recommend an immediate isolation of the installation and remedial action taken by a skilled person competent in such work.

Whereas, any defects revealed *during the Initial Verification process* must be rectified by the installer/constructor, having rectified the problem the installation or final circuit must be re-tested to confirm it meets the required standard before the Electrical Installation Certificate can be issued.

5.6 Required information

Before commencing with the inspecting and testing process the following information should be made available to the inspector:

• Earthing arrangements (see *Regulation 312*)
• Number and type of live conductors (see *Regulation 312*)
• Nature of supply parameters (see *Regulation 313*)
• Supply protective device characteristics (see *Regulation 313*).

Additional information in the form of *diagrams, charts or tables* should also be made available to the inspector in order that a safe inspection and testing can be conducted, which should include:

• The type and composition of each final circuit.
• The method used for basic and fault protection.
• Information necessary to identify the function and location of those devices which provide protection, isolation and switching.
• Identification of those circuits or equipment which are vulnerable to a particular test.

(See *BS 7671 Regulation 514.9.1 page 132.*)

101

SUMMARY

The establishment of a safe electrical installation is a systematic process: design and installation, with the Inspection and Testing via *Initial Verification*, being the final link and probably the most crucial. Consequently, the ultimate responsibility for the safe functioning of an installation rests with the inspector's *Initial Verification*; which can be summarised as follows:

1   **The Purpose of the Initial Verification** – to *confirm* an installation *complies with the design and construction standard required by BS 7671 Requirements for Electrical Installation.*

2   **Departure(s) from Regulation 120.3: Parts 3–7** – any *intended* departures from these *Regulations* will require special consideration by the designer, and they will:
    - be noted on the *Electrical Installation Certificate*
    - not affect the safety of the installation
    - not be less than that obtained by compliance with the *Regulations*. (Also see GN3 paragraph 1.5 page 14)

    If the *designer* has stipulated departures from *Regulations*, the inspector should:
    - request for the design criteria, or
    - forward the test results to the *designer* for verification
    - in the absence of the design criteria, *BS 7671 Regulations* will apply.

3   **Foremost Inspection** – this is not an intimate inspection of electrical equipment, more to confirm the equipment meets the current British and Harmonised standards, inasmuch as the electrical equipment has been selected and erected correctly, it is not damaged or defective. This procedure must take place before the commencement of any testing.

4   **Inspection of Installed Equipment** – this *is* an intimate inspection of the electrical equipment, for example: the relationship between conductors and protective devices or the requirements for RCDs.

5   **Inspector's Responsibilities** – to ensure the installation primarily meets BS 7671 standard, and conforms to the designer's criteria. To consider the safety of persons, livestock and not to cause damage to the property.

6   **Required Information** – before the inspector can begin any live testing he or she will need details of the DNO's (Distribution Network Operator) supply input and earthing arrangements.

7   **Additional information** – details of the installation are essential; the inspector will need to know, for example: the type and composition of each final circuit. Without this type of information it could possibly make the Initial Verification process extremely difficult or possibly inoperable.

Further related topic questions

Q4. An inspector has a number of <u>responsibilities</u> when carrying out an Initial Verification, the most significant is:

a) To ensure no danger occurs to any person and property is not damaged

b) To ensure no danger occurs to any person or livestock and property is not damaged.

c) To ensure the cable size (csa) is maintained throughout the cable length.

d) To check the designed cable sizing.

Answer: b) To ensure no danger occurs to any person or livestock and property is not damaged (see *GN3 Sections 1 & 2 paragraphs 1.2 & 2.1 pages 12 & 17*).

Reasoning

The inspector has four <u>responsibilities</u> when <u>inspecting</u> an installation, however the foremost is safety; that is, there is <u>no danger to persons or livestock</u>. It is *not* the responsibility of the inspector to check the designed cable size or whether it is maintained throughout the cable length. These responsibilities are those of the designer and installer respectively.

Q5. Information relating to the installation's <u>supply</u> should be readily available to the inspector; where the information is not forthcoming, which of the following cannot be achieved by calculation, measurement or inspection?

a) The installation's earthing arrangements

b) The overcurrent protective device at the origin of the installation

c) Number and type of live conductors

d) Nominal voltage and frequency.

Answer: d) Nominal voltage and frequency (see *BS 7671 Nature of Supply Parameters of the EIC page 463*).

Reasoning

The inspector needs to know, and understand, the requirements of the *Electrical Installation Certificate* (EIC), more so because this document may be used in a Court of Law and the inspector has *due diligence* to ensure the document is accurately completed (see the *Defence Regulation 29 of Electricity at Work Regulations 1989*).

Consequently, the reference to <u>voltage</u> and <u>frequency</u>, in the subsection: *Nature of Supply Parameters*, indicates these quantities <u>must be achieved by enquiry</u>.

Initial verification: scenario

The following electrical installation scenario has been reproduced courtesy of City & Guilds; however the multi-choice questions and the associated answers have been devised by the author.

An electrical installation within a new building has been completed and is about to be inspected and tested.

A five core thermoplastic 70°C steel wire armoured (SWA) 10 mm² cable is used to supply the new building and is fed from the main panel in an existing building and forms part of a three-phase 400/230 volts TN-S system. This cable has been installed in a 50 metre concrete service duct between the two buildings.

The values of Ze at origin of the installation, and its prospective fault current Ipf, are 0.12Ω. and 3.5 kA respectively.

All circuits within the new building are wired in 70°C thermoplastic single-core cable with copper conductors installed in steel conduit and trunking.

The metallic water installation pipe in the new building is bonded to the earthing terminal within the new building.

All testing is to be carried out in an ambient temperature of 20°C.

The conductor resistances in mΩ/m at 20°C.

Conductor csa (mm²)	mΩ/mat20°C
1.5	12.10
2.5	7.41
4.0	4.61
6.0	3.08
10	1.83

Q1. When inspecting the steel wiring of the armoured cable, which is used for mechanical protection, the inspector notes it is terminated in an earth terminal at the origin of the installation and in the new installation's DB. What action should the inspector take?
a) None, the swa is suitably earthed
b) Remove the swa from the earth terminal at the origin of the installation
c) Remove the swa from the earth terminal at the new installation's DB
d) Inform the constructor/installer of the problem.

Q2. When measuring the Zs at the new installation's DB, the inspector would expect to record an ohmic value of:
a) 0.12Ω
b) 1.83Ω
c) 0.303Ω
d) 0.063Ω

Q3. What Reference Method should the inspector record on the *Schedule of Test Results* for the distribution cable between the origin of the installation and the new DB?
a) A
b) B
c) C
d) D

Q4. The metallic water pipe in the new building is bonded to the earthing terminal within the new building; the inspector is to carry out a continuity test on the protective conductor, which is 30 metres in length with a csa of $6mm^2$. Give: the type of instrument to be used for the test, precautions to be taken to ensure accuracy of the test result, and the Test Method to be applied.
a) Low Impedance meter, null test leads and Test Method 1
b) Low Impedance Ohmmeter, calibration in date, null test leads and Test Method 1
c) Low Resistance meter, calibration in date, null test leads and Test Method 2
d) Low Resistance Ohmmeter, calibration in date, null test leads and Test Method 2.

Q5. Determine the ohmic value of the 30 metre, $6mm^2$ single-core cable.
a) 0.9Ω
b) 0.09Ω
c) 0.18Ω
d) 0.018Ω

Q6. The metal trunking of the new installation is to be inspected. What will the inspector be confirming?
a) The metal trunking is supplementary bonded, confirmed continuity and secure
b) The metal trunking joints are mechanically sound, earthed and secure
c) The metal trunking joints are mechanically sound, continuity, Line & Neutral are contained in the trunking, earthed and secure.
d) The metal trunking joints are mechanically sound, each circuit's conductors are contained in the trunking, earthed, adequate continuity and secure.

Q7. **What information should the inspector record on the Electrical Installation Certificate with regards to the bonding conductors?**

a) Location, material and verification

b) Connection, location and verification

c) Material, csa and connection

d) Material, length and connection.

Q8. **A prospective fault current test is to be carried out by the inspector on the installation. What is the purpose of this test?**

a) To ensure compliance with BS 7671 the test sequence

b) To ensure the distribution supply's protective device disconnects in 5 seconds

c) To ensure a circuit's protective device can safely interrupt the maximum fault current

d) To ensure a circuit's protective device can safely interrupt the minimum fault current.

Q9. **The inspector is required to determine the prospective fault current at *every relevant point of the installation*; with the initial reading taken at the origin of the installation what will the inspector observe when readings are taken at every *relevant point*?**

a) An increase in the value due to a decrease in the installation's resistance

b) An increase in the value due to an increase in the installation's resistance

c) A decrease in the value due to an increase in the installation's resistance

d) A decrease in the value due to a decrease in the installation's resistance.

Q10. **A phase rotating disc instrument is to be used by the inspector at the installed distribution board. What is the main reason for the inspector to carry out this test and where will the result be recorded?**

a) To ensure three phase motors rotate correctly. Readings are not recorded

b) To ensure the line conductors are correctly identified. EIC

c) To ensure the supply polarity is correct. EIC

d) To ensure the supply polarity is correct. EIC & Schedule of Test Results.

Answers and reasoning

Q1. Answer: d) Inform the constructor/installer of the problem

Reasoning: The fifth core, in the five core swa cable, will be used as the circuit's cpc, with the steel wiring of the armoured cable earthed at the origin of the installation and the new DB; a low impedance parallel path could be created, under given fault conditions, a circular fault path thereby preventing the distribution protective device from operating.

Although the inspector would probably remove the steel wiring from the DB's earth terminal it is, nevertheless, the responsibility of the constructor/installer to rectify any snags.

Q2. Answer: c) 0.303Ω

Reasoning: The inspector can, initially, determine the expected measured value of the earth fault loop impedance (Zs) with a simple calculation:

$$Zs = Ze + (R_1 + R_2) \times L$$

where $Z_e = 0.12\Omega$ at the origin of the installation

$R_1 = 1.83\Omega$ mΩ/m for the 10 mm^2 conductor (L1=L2 =L3)

$R_2 = 1.83\Omega$ mΩ/m for the cpc 10mm^2 conductor

L = 50m   the length of the concrete ducting neglecting the *tails* at the panel and DB

$$Zs = 0.12 + \left(\frac{1.83 + 1.83}{1,000}\right) \times 50$$

$$Zs = 0.12 + 0.183$$

$$Zs = 0.303\Omega$$

Where L1 is used in the initial measurement of Zs, the inspector would also take further measurements with:

- L2 (R$_1$) and cpc (R$_2$)
- L3 (R$_1$) and cpc (R$_2$)
- the highest measured value will be recorded on the Electrical Installation Certificate and the Schedule of Test Results.

Q3. Answer: d) D

Reasoning: The five core thermoplastic 70°C steel wire armoured (swa) 10 mm^2 cable has been installed in a 50-metre concrete service duct between the two buildings; therefore Appendix 4 of BS 7671 should be referred to, as follows:

- Paragraph 7.1 Reference Methods pages 383–384
- Table 4A2 page 390
- Table 4D4A page 407 Column 7.

Q4. Answer: d) Low Resistance Ohmmeter, calibration in date, null test leads and Test Method 2

Reasoning: To ensure the accuracy of the continuity reading, the instrument's calibration must be in date, the resistance of the instrument's leads must be *nulled* or alternatively their ohmic value subtracted from the ultimate test result and use Test Method 2 (wandering or long lead) with one end of the bonding conductor disconnected from the metallic pipe (see *GN3 pages 43–45*).

Q5. Answer: b) 0.09Ω

Reasoning: Length 30 metre

Resistance of cable 3.08 mΩ/m @ 20°C

$$30 \times 3.08/1000 = 0.0924\Omega$$

Answer: 0.09Ω (to two decimal places).

Q6. Answer: d) The metal trunking joints are mechanically sound, each circuit's conductors are contained in the trunking, earthed, adequate continuity and secure

Reasoning: Initially the inspector must confirm the trunking has been <u>securely</u> erected for obvious reasons.

The trunking must be <u>earthed to provide fault protection</u>; therefore all joints must be mechanically sound and a continuity test will verify this.

To <u>eliminate electromagnetic induction</u>, the line, neutral and cpc conductors for each final circuit must be contained within the metallic trunking (see *GN3 pages 37–38*).

Q7. Answer: c) Material, csa and connection

Reasoning: If the sub-section *Main Protective Bonding Conductors* of the EIC page 463 is reviewed, the inspector will record the following characteristics of the bonding conductor:

  i) Material (usually copper)
  ii) Cross sectional area (csa) in millimetres (mm^2)
  iii) Connection/continuity verified (with a tick ($\surd$) in the associated box).

Q8. Answer: c) To ensure a circuit's protective device can safely interrupt the maximum fault current

Reasoning: The purpose of the prospective fault current test is to ensure the maximum fault current does not exceed the Icn rating of the protective device; this value is usually given in a rectangular box on a circuit breaker, which ensures the circuit breaker can safely interrupt the maximum fault current (see *GN3 pages 68–70*).

Q9. Answer: c) A decrease in the value due to an increase in the installation's resistance

Reasoning: The installation's resistance will increase at each relevant point downstream of the origin. The main reason for this increase is the length(s) of the distribution cable(s) installed and their csa.

The application of Ohm's Law Equation ($I = V/R$) will assist in determining the prospective fault current; as the installation's resistance increases, and provided the supply voltage remain constant, the fault current will decrease.

Q10. Answer: d) To ensure the supply polarity is correct. EIC & Schedule of Test Results

Reasoning: BS 7671 Regulation 643.9 page 235 states: *for polyphase circuits, it shall be verified that the phase sequence is maintained*; in other words all three phase circuits must be tested to ensure the polarity is correct and this will be achieved using a phase rotating instrument.
(Also see *GN3 pages 73–74.*)

Initial verification: practical

All C&G 2391-50 candidates will carry out an Initial Verification on a pre-constructed installation fabricated to City & Guilds design and standard.

The candidate will be provided with a full description of the electrical installation to be inspected and tested, which should be read carefully – the data within the scenario is essential and necessary to carry out the visual Inspection and Testing of the installation.

The testing sequence must follow the one specified in BS 7671 and GN3; both documents may be referred to during the testing procedure to determine compliance.

BS 7671 Electrical Installation Certificate, Schedule of Inspection and Schedule of Test Results documentation will be provided.

Candidates may use their own test instruments; however it is their responsibility to ensure the accuracy of the instrument and compliance with the current Edition of GS 38. If the instrument, associated leads and probes fail to comply, the assessor may not permit their use during the assessment.

An open book, short answer question paper, relating to the practical concepts of the assessment will be completed by the candidate usually before commencement of the practical assignment.

The open book aspect of the question paper allows reference to the following documents: BS 7671, On-Site-Guide and GN3.

6
Periodic inspection

INTRODUCTION

Over a period of time an installation can be subjected to various adverse conditions, in addition to the accepted inherent deterioration related to the aging process, and general wear and tear associated with usage. Therefore it is a little unwise to leave an installation for long periods of time without some form of formal inspection.

The *Electricity at Work Regulations 1989* ensures the maintenance of commercial and industrial installations is not neglected. Consequently a formal inspection and testing arrangement must be implemented because it is effectively a statutory requirement; however if the installation is adequately supervised with continuous monitoring and planned maintenance, a formal Periodic Inspection need not be executed; whereas the condition of a domestic installation is essentially the responsibility of the occupier.

6.1 Purpose of periodic inspection and testing

What is the *reason* or *purpose* of a Periodic Inspection and Testing?

It is effectively an audit on an electrical installation; for the inspector to make a professional judgement on the installation's condition. Consequently the objective or purpose of a Periodic Inspection and Testing is to ensure that the installation is in a *satisfactory condition and can continue to be used safely*.

Whereas a Periodic Inspection and Testing should be carried out, so far as is reasonably practicable, to ensure:

- Persons and livestock are safeguarded against the effects of electric shock and burns.
- Property is protected against damage by fire and heat arising from an installation defect.

- Confirmation that the installation is not damaged or deteriorated so as to impair safety.
- Any defects, omissions and/or departures from *BS 7671 Regulations*, which may give rise to danger, are identified.

Initially the inspector should carry out a detailed visual examination of the installation to investigate the possible elements which could cause the installation to deteriorate, for example:

- general wear and tear
- damage through corrosion
- excessive electrical loading
- ageing and environmental influences.

On completion of the visual inspection, the inspector should conduct the appropriate tests which are conducted mainly to confirm the disconnection times given in BS 7671 *Chapter 41*, specifically: *Table 41.1, Regulations 411.3.2.3 and 411.3.2.6 page 59.*

6.2 Necessity for periodic inspection and testing

Over a period of time an electrical installation can deteriorate due to a number of factors (see Section 6.1), however there are other reasons why a Periodic Inspection and Testing is *necessary*. They are:

1 **Legal requirements** – legislation requires that electrical installations are maintained in a safe condition (see *Electricity at Work Regulations 1989*).

2 **Other interested organisations** – institutions which have a vested interest in a property or building such as insurance and/or mortgage companies or public bodies or licensing authorities.

3 **Compliance with BS 7671** *Requirements for Electrical Installation* – to confirm an installation is compliant with these *Regulations* when there has been:
- a change of occupancy of the premises
- a change of use of the premises
- any additions or alterations to the original installation
- a significant increase in the electrical loading of the installation
- possible damage to an installation, for example after flooding.

6.3 Required information

The *person who is responsible for the electrical installation* should be able to provide the inspector with the following:

- diagrams
- design criteria
- type of electricity supply
- earthing arrangements
- alternative supply (if applicable).

The *type of electricity supply and earthing arrangements* are recorded in *Section* I under *Supply Characteristics and Earthing Arrangements* of the *Electrical Installation Condition Report* (see Appendix 6).

- **Earthing arrangement** – limited to TN-C-S, TN-S & TT
- **Number and type of live conductors**
- **Nature of Supply Parameter** – *Nominal voltage and frequency* can only be ascertained by *enquiry only*, whereas the *prospective fault current and the external earth loop impedance* can be determined by either *enquiry or measurement*
- **Supply Protective Device** – the service or header fuse can be determined by either *enquiry or visual inspection* to establish the *BS (EN) number, the type and current rating.*

Whereas *legible diagrams, charts, tables or equivalent form of information* should conform to BS 7671 *Regulation 514.9.1*, specifying:

- the type and composition of each circuit (points of utilisation served, number and size of conductors, type of wiring)
- the method used for compliance with Regulation 410.3.2 (shock protection)
- the information necessary for the identification of each device performing the functions of protection, isolation and switching, and its location, and
- any circuit or equipment vulnerable to a typical test, as required by Part 6.

It is of the utmost importance that the inspector is informed, *before* any work commences, of the following:

- exactly what work is to be carried out? (*extent*)
- are there any areas which cannot to be inspected and tested, and the reasons? (*limitations*)

- if there are any restrictions who has given the authorisation? (*the agreed person*)

The *extent and limitations* of the inspection and testing are normally established with the *person ordering the work* or *the client* before the work begins, however the client or a representative may not be aware or fully understand the concept of *limitations* for whatever reasons. Therefore, consultations should take place to determine if there are any restrictions, which may limit the inspection and testing process. Having established the level of any limitations, they must be agreed with a *named person* and recorded on the *Electrical Installation Condition Report* in *Section D*.

6.4 What action should be taken where diagrams, charts or tables are not available?

A durable copy of a *schedule* incorporating detailed installation information is generally attached to the inside of a DB's door, however if the schedule is *not* available then the inspector should make a preliminary investigation of the installation to ascertain whether an inspection and testing can be carried out *safely and effectively*. Consequently, exploratory work will have to be conducted, which may include a *survey* to identify, for example, switchgear, controlgear and the circuits they control.

Where large and complex installations are involved, the lack of suitable data can create serious problems and possibly make a Periodic Inspection and Testing inoperable. Therefore the client should be advised of the problems and request the relevant information is made available before any work begins.

An alternative solution to this problem is for the inspector to conduct a limited inspection of the installation, provided the inspector considers it is safe to continue; however the unavailability of diagrams, charts or tables should be recorded on the *Electrical Installation Condition Report*.

6.5 What is a survey?

A survey gives an inspector an opportunity to gain a "feeling" or an impression of the installation, to assess its condition simply by reviewing the external and the internal conditions of the electrical equipment.

Warning bells should begin to ring if the survey reveals, for example: external damage to the equipment, signs of poor workmanship, exposed conductors or if supplementary bonding is missing or detached inside electrical control cabinets.

6.6 Sampling

A simple *walk-round survey* will assist an inspector to establish the proportion or *sample size* of the installation he or she will inspect and test. If, during the Inspection and Testing procedure, the initial sample size highlights defects or deficiencies, then the sample size needs to be increased.

A *sampled size should* only be a *representative proportion* of the installation, *not* the whole installation simply because a 100% inspection and/or testing is:

- unrealistic
- uneconomical
- unachievable in many installations because of their physical size.

Note: these restrictions do not apply to domestic and less complex installations.

6.7 Setting the sample size

Determining how large a sample size should be, or what proportion of an installation should be inspected and tested, can be problematic or at least challenging for an inspector.

The initial sample size can depend upon a number of factors, not least the experience of the inspector; however one or more of the following factors could assist:

- the physical size of the installation
- approximate age and condition
- type and usage
- ambient conditions
- maintenance
- previous inspection/testing details
- consultation with the owner
- quality of records such as electrical installation and minor works certificates, previous Periodic Inspection reports and maintenance records.

Further guidance on sample sizes, and the range of items to be inspected, is given in *IET's Guidance Note 3 Table 3.3*, although by rule of thumb it is recommended that at least 10% of each section selected is sampled. Irrespective of the sample size, *all earthing and protective bonding conductors* must be checked to ensure they are *present and secure*.

6.8 Results of sampled inspection and testing

If the initial sampled size produces poor or unacceptable results, then the inspector could draw the conclusion that there is a reasonable probability that similar problems may exist within the remaining section not inspected and/or tested.

Therefore the options open to the inspector are to:

* increase the sample size
* refer back to the client
* recommend a full (100%) Inspection and Testing of the selected section initially sampled but *not* the complete installation!

Related topic questions

Q1. State the purpose of a Periodic Inspection and Testing.

a) A client has requested a Periodic Inspection and Testing

b) A Periodic Inspection and Testing is a legal requirement

c) An installation is in a satisfactory condition and can continue to be used safely

d) To ensure the installation complies with BS 7671.

Answer: c) An installation is in a *satisfactory condition and can continue to be used safely*.

Reasoning

The question requires the purpose of a Periodic Inspection and Testing. Consequently the inspector must ensure the installation is in a satisfactory condition; moreover it can continue to be used safely.

Do not forget: a periodic inspection is carried out on an active installation, subsequently why should this type of inspection and testing be carried out? The answer is the one given: to ensure the installation is fit for purpose, inasmuch as it is in a satisfactory condition and can continue to be use safely.

Q2. An electrical installation in an industrial unit is scheduled for a <u>Periodic Inspection</u>. There are *no* previous documents or circuit charts available. What action should the inspector take?

 a) Carry out a complete Periodic Inspection and Testing of the industrial unit
 b) Do not to carry-out the Periodic Inspection and Testing until the relevant documentation is produced
 c) Carry out an exploratory survey
 d) Liaise with the client before carrying out an exploratory survey.

Answer: d) Liaise with the client before carrying out an exploratory survey

Reasoning

The size of the installation is not given, therefore an *exploratory survey* is a realistic response. A full or 100% inspection and testing is *not* acceptable; however, the inspector needs to consult with the client before any work is carried out in order to comply with Section D of the EICR.

Note: City and Guild's Chief Examiner has previously made some critical comments regarding the poor comprehension many candidates have when asked *what action should the inspector take <u>before</u> conducting an inspection and testing when there were no previous documents, details or circuit charts available to him or her?* Items 6.4 to 6.8 can assist the candidate overcome these criticisms.

(Also see GN 3 Section 3.8.3 page 89 and BS 7671 Regulation 651.1 page 237.)

Q3. When is it not necessary to carry-out a Periodic Inspection of an installation?

 a) There has been a change of occupancy of the premises
 b) There has been a change of use of the premises
 c) There have been new additions and alterations to the installation
 d) There has been flood damage to the installation.

Answer: c) There have been new additions and alterations to the installation

Reasoning

Any <u>new additions or alterations</u> to the original installation will require either initial verification or the issue of a *Minor Electrical Installation Works Certificate*, depending on the type of additions or alteration.

Q4. What is the purpose of a survey before a Periodic Inspection can be carried out?

 a) To confirm an installation is in a satisfactory condition
 b) To help plan the inspection and the size of the inspection sample
 c) To ensure the installation is safe and can continue to be used
 d) To ensure the installation meets the standard of BS 7671.

Answer: b) To help plan the inspection and the size of the inspection sample

6.9 Frequency of periodic inspections

The time interval between the *initial verification* and the *first Periodic Inspection* is recommended by the installation's designer(s). Thereafter, the responsibility for subsequent inspection(s) is the prerogative of the *skilled electrical person* who is *competent to carry out inspection and testing* (see BS 7671 Regulation 652.1 page 237).

There are legal requirements that stipulate time intervals for *routine checks* and the *maximum period between each inspection and testing*; there are also other reasons for an inspection and test, for example, a prerequisite for a mortgage or insurance (see Section 6.2).

Nevertheless, the inspector must also make a professional judgement when recommending the next Periodic Inspection based upon, for example the following:

- the type of installation
- the equipment, its usage and operation
- any known maintenance
- external influences
- general condition, any signs damage or deterioration.

6.10 Competence

The ability and competency of a skilled person to carry out Periodic Inspection and Testing is paramount; therefore it will be expected that an inspector has had a formal education to successfully achieve electrical installation qualifications, training and practical skills; to have gained experience and knowledge to enable him or her to be fully conversant with those aspects required to carry out the critical task of inspection and testing.

SUMMARY

An inspector must have a depth of knowledge and experience to enable him or her to carry out the practical aspects of a Periodic Inspection and Testing. However, as a C&G 2391-51 and 52 candidate, the inspector must also understand why a Periodic Inspection is necessary. Accordingly, he or she must be familiar with the following:

- **The purpose of Periodic Inspection and Testing**: to determine whether or not an installation is in a *satisfactory condition* where it can *continue to be used in a safe way*.
- The objectives, when conducting a Periodic Inspection and Testing, are to ensure:
    - i) persons and livestock are safeguarded against the effects of electric shock and burns
    - ii) property is protected against damage by fire and heat arising from an installation defect
    - iii) the installation is not damaged or has not deteriorated so as to impair safety
    - iv) any defects, omissions and/or departures from *BS 7671 Regulations*, which may give rise to danger, are identified.
- **Why Periodic Inspections and Testing are necessary**: for legal reasons and to ensure installations, under given circumstances, comply with BS 7671.
- **Why is installation information required by an inspector?** To assist the inspector in ascertaining whether an installation is in a satisfactory condition and safe to continue to use with the aid of:
    - i) diagrams
    - ii) design criteria
    - iii) type of electricity supply
    - iv) earthing arrangements
    - v) alternative supply.
- **What action should the inspector take when the information is _not_ available?** This will depend upon given circumstances. Initially a propriety investigation, usually via an inspection, of the installation should take place before a Periodic Inspection and Testing is carried out. Provided the inspector is satisfied with the investigatory results, a survey can be conducted.
- **What is a survey?** Initially a *walk round* the installation to inspect the condition of the electrical equipment, to help form an opinion on the amount or the *sample size* of the installation to be inspected and tested.

The inspector will need, for example, to carry out an inspection of the installation's switchgear, to be able to determine the age of installation components and to recognise signs of deterioration.

The external condition of the electrical equipment and wiring systems generally gives a good indication of the standard of maintenance and any signs of deterioration, typically:

i)   safety – IP Codes, equipment earthing
ii) age – old Vulcanized India rubber cables, insulation perishes, cracks and falls off, no cpc
iii) damage – cables with exposed conductors
iv) corrosion and external influence – metal conduit corroded, no EMI protection
v)   sign of overloading – cables discoloured
vi) wear and tear and environment – general condition, is the wiring system suitable for the environment?
vii) suitability for continued use – unsatisfactory if C1, C2 or FI codes are issued.

- What proportion or sample size of the installation should be inspected? Approximately 10% of a selected section, but *not* 100% of the installation?
- How frequently should a Periodic Inspection take place? There are legal reasons for the time intervals between Periodic Inspections. However the inspector must also make a professional judgement when recommending the next Periodic Inspection, which could depend on the elements highlighted in Section 6.9.

Further related topic questions

Q5. A durable copy of a <u>schedule</u>, relating to a distribution board (DB), shall be provided within the DB or adjacent to it. The schedule should be made available to the inspector, which of the following must be recorded on the schedule:
   a) The installation's designer
   b) The installation's installer
   c) The installation's shock protection
   d) The installation's compliance with Regulation 410.3.2.

Answer: d) The installation's compliance with Regulation 410.3.2 (see *BS 7671 Regulation 514.9.1 (ii) page 132*).

Reasoning

A _schedule_ should give a comprehensive description of the installation, ranging from the number of conductors to electrical items which are vulnerable to tests.

To ensure the _schedule_ (and the installation) complies with Regulation 514.9.1, the inspector must refer to Regulation 410.3.2 for compliance.

Q6. A Periodic Inspection is required for a large industrial installation; there are no diagrams, charts or tables available to the inspector. What action should the inspector not carry out?

a) Conduct an exploratory survey to enable the inspection and testing to be carried out safely and effectively

b) Limit the Periodic Inspection to a visual inspection

c) Advise the client that diagrams, charts or tables must be made available before the inspection and testing can take place

d) Carry out the Periodic Inspection and record on the EICR that diagrams, charts or tables were not available.

Answer: d) Carry out the Periodic Inspection and record on the EICR that diagrams, charts or tables were not available (see _IET Guidance Note 3 paragraph 3.8.3 General Procedure_).

Reasoning

There are a number of options available to the inspector, as highlighted in GN3, in situations where diagrams, charts or tables are not available, but simply recording on the EICR that the relevant documentation was not available is not one them.

Periodic inspection: Scenario

The following electrical installation scenario has been reproduced courtesy of City & Guilds; however the multiple-choice questions and associated answers have been devised by the author.

A client has requested that their 7-year-old, small commercial unit has a Periodic Inspection and Test for insurance purposes.

The installation forms part of a 400/230 V, three-phase, TN-C-S system having a previously recorded Ze and Ipf of 0.25Ω and 2.0 kA respectively.

There have been a number of alterations to the original installation, which are supported with Minor Electrical Installation Works Certificates.

All final circuits are installed using 70°C thermoplastic single-core cables, with copper conductors, installed in metal conduit and trunking throughout.

The protective devices within the installation are a mixture of BS EN 60898 circuit breakers and BS EN 61009 RCBOs, all having an I_{cn} rating of 6 kA.

All testing is carried out at a temperature of 20°C.

Cable csa: – Lighting 1.5 mm^2

Ring Final Circuit 2.5 mm^2

Lathe and radial power 4.0 mm^2

Figure 6.1

Way	Description	Protective Device	Ring Final Circuit Continuity (Ω)				Insulation Resistance (MΩ)	Zs (Ω)	RCD IΔn
			r_1	rn	r_2	(R1 + R2)			
1L1	Office Power	B32	0.8	0.7	0.8	0.62	0.5	1.2	310
1L2	Workshop Power	B32	0.75	0.95	0.8	1.9	250	0.9	210
1L3	Water Heater 3kW	B16	–	–	–	1.2	≥500	1.5	N/A
2L1	3 Phase Lathe	C16	–	–	–	0.8	350	0.85	N/A
2L2	3 Phase Lathe	C16	–	–	–	0.75	300	0.9	N/A
2L3	3 Phase Lathe	C16	–	–	–	0.85	400	0.95	N/A
3L1	Office Lighting	C10	–	–	–	0.44	≥500	0.69	N/A
3L2	Workshop Radial	B10	–	–	–	1.35	≥500	1.55	N/A
3L3	Warehouse Lighting	C10	–	–	–	2.5	≥500	2.8	N/A
4L1	Spare	–	–	–	–	–	–	–	–
4L2	Spare	–	–	–	–	–	–	–	–
4L3	Spare	–	–	–	–	–	–	–	–

Q1. **Give the primary reason for the Periodic Inspection.**
   a) The age of the installation
   b) The age of the installation and insurance purposes
   c) Insurance purposes
   d) It is a commercial installation.

Q2. **The combined resistance of the of circuit 1L3 is 0.44Ω at 20°C and the load current is 13Amperes. Determine the voltage drop for the circuit: does this calculated value comply with BS 7671 and what is the maximum voltage drop for 1L3?**
   a) Calculated: 5.72 volts, Does not comply with BS7671, Maximum voltage 7.04 volts

b) Calculated: 5.72 volts, Does comply with BS7671, Maximum voltage 5.72 volts

c) Calculated: 5.72 volts, Does comply with BS7671, Maximum voltage 11.5 volts

d) Calculated: 5.72 volts, Does comply with BS7671, Maximum voltage 7.04 volts.

Q3. An insulation resistance test is to be carried-out on Circuit 3L1. State the instrument to be used for the test, the maximum test voltage and the Observation Code to be recorded if the insulation resistance reading is half the minimum value given in BS 7671.

a) Insulation resistance ohmmeter, 250 volts dc and C1

b) Insulation resistance ohmmeter, 500 volts dc and C2

c) Insulation Resistance Tester, 500 volts rms and C2

d) Insulation Resistance Tester, 500 volts dc and C2.

Q4. What action should the inspector take regarding the measured earth fault loop impedance (Zs(m)) readings?

a) Compare the measured values with those given in Table 41.3 Provided the values are less or equal no further action is necessary

b) Apply the correction factor given in Appendix 3 to the measured values. Provided the values are less or equal no further action is necessary

c) Apply the correction factor to the values given in Table 41.3. Compare the value with measured value. Provided the measured values are less or equal no further action is necessary

d) Apply the correction factor to the values given in Table 41.3 and compare the results with the measured values. No further action is necessary.

Q5. What action should the inspector take regarding 3L2, where the design criteria identify the combined resistance of this circuit as 0.92Ω?

a) Circuit appears to be satisfactory with an insulation reading of ≥ 500MΩ

b) Investigate the reasons why the external earth loop impedance exceeds the correct value given in Table 41.3

c) Issue appropriate Observation Code 3 for the exceeded values of $R_1 + R_2$ Installation Satisfactory

d) Issue an Observation Code 2 for the exceeded values of $R_1 + R_2$ and a Code 3 for no RCD Installation Unsatisfactory.

Q6. When carrying out Step 3, continuity test of 1L2 Workshop Ring Final circuit, the inspector records a large increase in the ohmic value from one socket outlet to the next one; at approximately the mid-point of the ring final circuit there is a significant drop in the ohmic value from one socket outlet to the next one as the test continues towards the origin of the circuit. What action should the inspector take?

a) Assume it is the effect of parallel earth paths

b) Check the conductor's termination at each socket outlet

c) Assume there are a number of spurs in the ring final circuit

d) Transpose the open-ended conductors cross connected in the test terminal block.

Q7. During the inspection process corrosion is found on the top flat surface of the metallic trunking. What action should the inspector take?

a) Take no action

b) Look for the source of the problem

c) Issue an Observation Code C2

d) Issue an Observation Code FI.

Q8. The inspection of cables in metallic trunking system revealed IT cables. What action should the inspector take?

a) Check the distance between the LV and the IT cables is ≥ 130 mm

b) Check the distance between the LV and the IT cables is ≥ 150 mm

c) Check the distance between the LV and the IT cables is ≥ 200 mm

d) Check the distance between the LV and the IT cables is $= 0$ mm.

Q9. What action should the inspector take to ensure the safety of the installation has not been compromised by the alterations to the original installation?

a) Inspect the Minor Electrical Installation Works Certificates

b) Inspect the alterations

c) Inspect the alterations to ensure the earthing and bonding are adequate

d) Inspect the installation details for the alterations.

Q10. What is the correct testing method the inspector should adopt for measuring the earth fault loop impedance of the lathe circuit?

a) Live test, measurement taken at the load side of the lathe motor. Continuity of the motor's cpc measured separately

b) Live test, measurement taken at the supply side of the motor. Continuity of the motor's cpc measured separately

c) Live test, measurement taken at the supply side of the isolated motor

d) Live test, measurement taken at the supply side of the isolated motor. Continuity of the motor's cpc measured separately.

Answers and reasoning

Q1. Answer: c) Insurance purposes

Reasoning: The scenario clearly states the Periodic Inspection is for <u>insurance purposes</u>. Although there could be a secondary reason; that is, the age of the installation, this reason is not given in the scenario and the candidate must not make any assumptions.

Q2. Answer: d) Calculated: 5.72 volts, Does comply with BS7671, Maximum voltage 7.04 volts

Reasoning: With the aid of Ohms Law Equation $V = I \times R$.

Where V is the voltage drop,

I is the load current of 13A, and

R is the resistance of both the line and neutral conductors 0.44Ω.

$V_d = 13 \times 0.44 = 5.72$ volts.

Regulation 525 page 145 *Voltage Drop in Consumers Installations* refers to Appendix 4, Section 6.4, Table 4Ab page 383.

Table 4Ab refers to: "lighting" and "Other uses"; the latter applies to circuit 1L3.

The maximum permissible voltage for "Other uses" is 5% of 230v = 11.5 voltage. The calculated value of 5.72 volts is significantly less than the maximum permissible voltage therefore it is acceptable.

The maximum current circuit 1L3 can draw is 16A ($I_n = I_b$ Regulation 433.1.1 (i) page 89) $V_d = I \times R = 16 \times 0.44 = 7.04$ volts, which is less than the maximum of 11.5 volts.

Q3. Answer: d) Insulation Resistance Tester, 500 volts dc and C2

Reasoning: see BS 7671 Regulation 643.3.2 Table 64 page 232 and GN 3 paragraph 2.6.7 pages 49–53 where the instrument type and test voltage are given. The Observation Code can prove to be problematic: does the poor insulation reading gives rise to immediate danger (C1) or is it potentially dangerous (C2)?

The installation is 7 years old and in constant use, therefore it is unlikely that there is an immediate danger; more likely circuit 3L1 is potentially dangerous. Irrespective, the installation must be recorded as unsatisfactory.

Q4. Answer: c) Apply the correction factor to the values given in Table 41.3. Compare the value with measured value, provided the measured values are less or equal no further action is necessary.

Reasoning: The measured value, Zs(m), is taken at 20°C whereas the values given in Table 41.3 page 62 are calculated at 30°C, therefore the correction factor of 0.8 (Appendix 3) must be applied to those values given in Table 41.3.

Provided the corrected value is <u>not</u> greater than the measured value no action is necessary, however where the measured values are greater than the corrected values further action is necessary with an appropriate Observation Code being recorded (see *GN3 paragraph 2.6.16 item (d) page 68 and On-Site-Guide Appendix I Table 11 page 196*).

Q5. Answer: d) Issue an Observation Code 2 for the exceeded values of $R_1 + R_2$ and a Code 3 for no RCD Installation Unsatisfactory.

Reasoning: The original installation was installed under previous Regulations (installed 7 years ago), where socket outlets, not exceeding 20A, did not require RCD protection in commercial installations. Therefore an Observation Code

3 should be issued to recommend the fitting of an RCD to provide additional electrical safety.

There appears to a significant deterioration in the combined resistance ($R_1 + R_2$) of the circuit, which could present a potentially dangerous condition Therefore an Observation Code 2 should be issued making the installation unsatisfactory.

Q6. Answer: d) Transpose the open-ended conductors cross connected in the test terminal block.

Reasoning: With single core cables, care must be taken to ensure the opposite ends of the open-ended conductors are cross connected to achieve the objective of Step 3 ($R_1 + R_2$).

Q7. Answer: d) Issue an Observation Code FI.

Reasoning: A major responsibility for the inspector is to ensure the installation is safe to continue to be used, not to fault find. The source of the corrosion is not known or if it has caused any damage to the wiring system; therefore further investigations (FI) would be necessary to determine whether the installation is safe to continue to be used.

Q8. Answer: d) Checking the distance between the LV and the IT cables is = 0 mm.

Reasoning: It is not necessary to segregate LV and IT cables in metallic trunking because the metallic enclosure will effectively create screening from magnetic induction (see *BS 7671 Table A 444.1 page 119*).

Q9. Answer: c) Inspect the alterations to ensure the earthing and bonding are adequate.

Reasoning: The earthing and bonding for the alterations should have been installed to BS 7671 standard, by the constructor named on the Minor Works Certificate. However it is the responsibility of the inspector to ensure the all the protective measures are adequate (see *BS 7671 Regulation 132.16 page 21*).

Q10. Answer: d) Live test, measurement taken at the supply side of the isolated motor Continuity of motor's cpc measured separately.

Reasoning: The measurement of the earth fault loop impedance must be taken on the supply side of the lathe's isolated motor control gear for safety and logical reasons: if the measurement were taken at the motor it would require the motor to be switched on. The net result would be the creation of a dangerous and an unsafe situation.

The continuity of the protective conductor, from the motor to the isolated control gear, needs to be confirmed for safety reason when the lathe is operational.

(See *GN3 Motor Circuits page 100.*)

Periodic inspection: Practical

All C&G 2391-51 candidates will carry-out a Periodic Inspection on a pre-constructed installation fabricated to City & Guilds design and standard.

The candidate will be provided with a full description of the electrical installation to be inspected and tested, which should be read carefully The data within the scenario is essential and necessary to carry out the visual inspection and testing of the installation.

The testing sequence must follow the one specified in BS 7671 and GN3. Both documents may be referred to during *the testing procedure* to determine compliance; however they *may not be used* when carrying out the visual inspection component of the assessment.

The visual inspection will also require a candidate to examine a number of photographs displaying components or locations. The candidate must determine whether the item(s) displayed is (are) either compliant or non-compliant with BS 7671. The candidate will *not* be required to support their decision with a given Regulation; although there will be a requirement to allocate an Observation Code with a written Report on any departure(s) from the Regulations. Furthermore, any comments made in the Report must be based on the situation shown in the photograph; no assumptions should made.

Example:

Photograph no.	Non-compliance	Possible consequence	Observation code
56	Live Conductor exposed on motor circuit	Immediate danger	C1

BS 7671 Electrical Installation Condition Report, Schedule of Inspection and Schedule of Test Results documentation will be provided.

Candidates may use their own test instruments; however it is their responsibility to ensure the accuracy of the instrument and compliance with the current Edition of GS 38. If the instrument, associated leads and probes fail to comply, the assessor may not permit their use during the assessment.

An open book, short answer question paper, relating to the practical concepts of the assessment will be completed by the candidate usually before commencing with the practical assignment.

The open book aspect of the question paper allows reference to the following documents: BS 7671, On-Site-Guide and GN3.

7

Initial verification and periodic inspection questions

The C&G 2391-52 is, as the title intimates, a combination of both types of inspection and testing; therefore it tends to address those candidates who possibly have significant experience/background within the inspecting and testing environment.

The format for this Unit is similar to the other two Units insomuch as it is an open book, on-line and multiple-choice tests.

The candidate will be presented with an installation scenario from which multiple-choice questions will be asked; plus a number of general multiple-choice questions.

The practical format for C&G 2391-52 is a combination of C&G 2391-50 & 51 given in Chapters 4 and 5 and will consist of three Tasks: A, B & C, which the candidate will have to complete. They are:

- **Task A**: Visual Inspection: Time allowed: 30 mins.
- **Task B**: Combined Initial Verification and Periodic Inspection of an Electrical Installation: Time allowed: 2 hours 30 mins.
- **Task C**: Short Answer Questions. Time allowed: 1 hour 20 mins.

Once again the candidate is advised to read the question and the multiple-choice answers carefully before making a selected response when completing the theory component. Nevertheless the candidate can always review the questions and revise the answer if necessary because the exam is on-line; whereas the practical is *pen & paper*.

Initial verification and periodic inspection: scenario

The following Scenario has been reproduced courtesy of City & Guilds; however the multiple-choice questions and associated answers have been devised by the author.

The electrical installation in a 15-year-old village hall requires inspection for a local authority license to hold functions. The building is used for various group activities each morning during the week and this is to continue during the inspection and test.

The supply and installation form part of a single-phase 230 volt TN-S system having a Ze of 0.03Ω.

All circuits are installed using thermoplastic 70°C single core cables, having copper conductors enclosed in surface mounted PVC conduit and trunking. The cpcs are the same csa as the live conductors throughout.

One additional ring final circuit was installed approximately 5 years ago to supply socket-outlets in the stage area. There is no evidence of any other alterations or additions to the installation. The certification from the Initial Verification of the original installation and suitable circuit charts are available for the inspector. There is no certification available for the additional ring final circuit.

Metallic oil and water installation pipework is installed within the building and 10 mm² main protective bonding conductors are installed within the building fabric and connected to the pipework.

A schedule, in a plastic envelope, attached to the inside of a metal-clad distribution board door, complies with *Regulation 514.9.1* and highlights the type of protective devices, which are a mixture of *Type B, BS EN 60898* circuit breakers and *BS EN 61009 RCBOs*.

All testing is carried out at a temperature of 20°C.

Figure 7.1

Circuit No.	Device rating	Description	Conductor csa in mm² Live	cpc	Cable Length (m)	Insulation Resistance (MΩ)
1	32A	Ring final circuit for socket-outlets main hall	2.5	2.5	50	200
2	32A	Ring final circuit for socket-outlets other areas	2.5	2.5	50	150
3	32A	Cooker	6.0	6.0	50	200
4	16 A	Immersion heater	2.5	2.5	80	50
5	16 A	Boiler	2.5	2.5	50	50
6	10A	Lights main hall	1.5	1.5	60	200
7	10A	Lights other areas	1.5	1.5	60	25
8	6A	Outside lights	1.5	1.5	80	1.5
9	32A*	Ring final circuit stage area	2.5	2.5	50	200
10	–	Spare	–	–	–	–

* Indicates BS EN 61009 type B RCBO

Questions

Q1. Which document must be completed by the inspector for the inspection and testing of the original installation?

a) Electrical Installation Certificate

b) Electrical Installation Minor Works Certificate

c) Electrical Installation Condition Certificate

d) Electrical Installation Condition Report.

Q2. What should the inspector discuss with the person ordering the inspection and testing before the work takes place?

a) The time and date when the work will begin

b) Extent and limitation of the work and any remedial work

c) What will be covered by the Report and the fee for the work

d) Extent and Limitation of the Inspection and Testing.

Q3. Give the reason why the village hall's electrical installation requires an inspection.

a) The age of installation

b) Change of use of the village hall

c) It is a statutory requirement

d) To make an application to the local authority to hold functions.

Q4. Determine the volts drop for Circuit 1 and whether it is acceptable. Assume the design current (Ib) for the ring final circuit, at the mid-point of the circuit, is half the maximum current for Circuit 1's wiring system (½Ib).

a) 12.2 volts; it is not acceptable

b) 12.2 volts; it is acceptable

c) 10.8 volts; it is not acceptable

d) 10.8 volts; it is acceptable.

Q5. Determine the maximum permissible measured value for Circuit 1's earth fault loop impedance to two decimal places.

a) 1.37Ω

b) 1.096Ω

c) 1.10Ω

d) 1.09Ω

Q6. Determine the earth fault loop impedance for Circuit 3.

a) 0.762Ω

b) 0.338Ω

c) 0.492Ω

d) 0.261Ω

Q7. Give the number of documents on which the value of external earth fault loop impedance will be recorded on for the complete installation.

a) 1

b) 2

c) 3

d) 4

Q8. Which boxes in Section I should the inspector tick for the electrical installation's supply?

a) AC and polarity

b) AC and 2 phase 3 wire

c) AC and 1 phase 2 wire

d) AC supply polarity and 1 phase 2 wire.

Q9. The short circuit capacity (Icn) for Circuit 1 is 6kA. Is this breaking capacity suitable for the protective device? If not, what action should the inspector take?

a) A short circuit braking capacity of 6kA is suitable for domestic installations; no action is required

b) Icn and Ics values are rated equal; no action is required

c) The highest measured short circuit current is greater than Icn; an Observation Code should be raised

d) The highest measured short circuit current is greater than Icn; an Observation Code should not be raised.

Q10. State the number of electrical tests that the inspector must carry out on an additional protection device.

a) 2

b) 4

c) 3

d) 6

Q11. State the function of the test button on the additional protection device.

a) To confirm the continuity of the installation's earth

b) To test the sensitivity of the device

c) To confirm the continuity of the circuit's cpc

d) The mechanical function of the device.

Q12. When carrying-out either an RCD or Earth Fault Loop Impedance test. What is the acceptable current rating of the probe's protective fuse?

a) 500mA HRC

b) 10A

c) 5A

d) 50mA HBC.

Q13. What action should the inspector take regarding the additional ring final circuit for the stage area?

a) Request the person ordering the work to supply further data on the additional circuit

b) Include the circuit in the inspecting and testing procedure: the circuit was constructed 5 years ago

c) Carry out a visual inspection and test the circuit: no documentation available

d) Carry out an Initial Verification; confirm the circuit complies with the current wiring Regulations.

Q14. Give the first three instruments tests, and the associated instrument used, when testing the electrical installation.

a) Continuity of protective conductors, continuity of ring final circuits and insulation resistance: continuity tester and insulation tester

b) Continuity of protective conductors, continuity of ring final circuits and insulation resistance: continuity tester and insulation resistivity tester

c) Continuity of protective conductors including main and supplementary bonding, continuity of ring final circuit conductors and insulation resistance: Low-resistance ohmmeter and insulation resistance meter

d) Continuity of protective conductors including bonding conductors and, continuity of ring final circuit conductors, insulation resistance and SELV protection: Low-resistance ohmmeter and insulation resistance tester.

Q15. 10 mm^2 main protective bonding conductors have been installed and connected to the metallic oil and water installation pipework. Give the recommended csa of the main earth conductor.

a) 10 mm^2

b) 6 mm^2

c) 16 mm^2

d) 25 mm^2

Q16. What action should the inspector take when the measured prospective fault current exceeds the Ics rating of a protective device but not the Icn rating of the circuit breaker?

a) No action because the device can interrupt a fault current safely without loss of performance

b) Take action because the device cannot interrupt a fault current safely without loss of performance

c) Take action because the device cannot interrupt a fault current safely without loss of performance. Issue an Observation Code C2

d) Take action because the device can interrupt safely without a loss of performance. Issue an Observation Code C3.

Q17. A number of warning notices, *of such durable material to remain of legible throughout the life time of the installation*, should be fixed in a prominent position near the origin of the installation. Give the items which are identified, and conform, to those given Section 4: Consumer Unit(s)/Distribution Board(s) of the *Condition Report Inspection Schedule*.

a) RCD Quarterly, non-standard cables and alternative supply

b) RCD three monthly, non-standard cables and inspection and testing

c) RCD Quarterly, alternative supply and non-standard cables

d) RCD six monthly, non-standard cables and Periodic Inspection and Testing.

Q18. The IP rating of the installation's distribution board enclosure must conform to Item 4.3 in Section 4: Consumer Unit(s)/Distribution Board(s) of the *Condition Report Inspection Schedule* which are:

   a) IP2X
   b) IP4X or IPXXB and IP2X or IPXXD
   c) IP4X
   d) IP2X or IPXXB and IP4X or IPXXD.

Q19. Figure 1 gives individual circuit insulation resistance readings. What would be the overall measured insulation resistance value?

   a) 1.5 MΩ
   b) 1.29 MΩ
   c) 2.5 MΩ
   d) 150 MΩ

Q20. A detailed inspection/examination with appropriate testing is required when conducting a Periodic Inspection and Testing. The inspection is made without dismantling equipment, as far as is reasonably practical, and the testing is to confirm:

   a) The test results are satisfactory
   b) The test results meet the designer criteria
   c) The disconnection times meet those stated in Chapter 41
   d) The disconnection times meet those stated in Chapter 41 and highlight other defects.

Answers and reasoning

Q1. Answer: a) Electrical Installation Condition Report.

Reasoning: An Electrical Installation Condition Report, with appropriate Schedules, must be completed when carrying out a Periodic Inspection and Testing.

Q2. Answer: d) Extent and Limitation of the Inspection and Testing.

Reasoning: See Section D of the Electrical Installation Condition Report (BS 7671 page 473).

Q3. Answer: c) It is a statutory requirement.

Reasoning: The answer to this question could be either c) or d), however the former is correct because it is supported by Regulation 115.1 page 17 and Item 6 in Appendix 2 page 360 of BS 7671; whereas the latter is the reason why the statutory requirement is necessary.

Q4. Answer: d) 10.8 volts; it is acceptable.

Reasoning: Volts Drop = (mV/A/m) × Ib × L

   Where (mV/A/m) is 18mV/A/m
         Ib is 12 Amperes
   and L is 50 metres.

The wiring system is <u>surface mounted, PVC conduit and trunking</u> with <u>single core cables</u> enclosed, which is Reference Method B.

Therefore the inspector should refer to Table 4D1A, page 401, and correlate Columns 1 and 4 for the maximum current for 2.5 mm^2 cable conductors which is 24 Amperes (Ib), but only half this current is required for this calculation.

The value of (mV/A/m) can be determined with the aid of Columns 1 and 3 on page 402 where the value is given as: 18mV per ampere per metre.

Volts Drop = 18/1000 × 12 × 50 = 10.8 volts.

Q5. Answer: c) 1.10Ω

Reasoning: The <u>maximum</u> Zs for a Type B circuit breaker is 1.37Ω (see Table 41.3 page 62) however the question required the <u>maximum measured value to two decimal places</u>. The Zs values given in Table 41.3 are calculated for the optimal ambient temperature of 30°C and the conductor operating temperature of 70°C; therefore the correction factor of 0.8 must be applied when measuring the earth fault loop impedance for a final circuit.

This achieved as follows: 1.37 x 0.8 = 1.096Ω, however the question requires a value to two decimal places; consequently the calculated answer must be rounded-up to:

Zs(m) = 1.10 Ω

(see GN3 2.6.16 item (d) page 68 and Appendix A2 page 143)

Q6. Answer: b) 0.338Ω

Reasoning: Use the formulae: Zs = Ze + (R$_1$ + R$_2$) × L

where Ze = 0.03

(R$_1$ + R$_2$) = 6.16 mΩ/m (see GN 3 page 148)

and L = 50m

Zs = 0.03 + (6.16 × 50)/1000

Zs = 0.03 + 0.308

Zs = 0.338Ω

Q7. Answer: d) 4

Reasoning: The inspector is required to complete the following documents for the complete installation:

i) Electrical Installation Certificate plus a Schedule of Inspection and a Schedule of Test Results.

ii) Electrical Installation Condition Report plus a Schedule of Inspection and a Schedule of Test Results.

The external earth fault loop impedance (Ze) is recorded on the following:

i) Electrical Installation Certificate and the Schedule of Test Results.

ii) Electrical Installation Condition Report and the Schedule of Test Results.

iii) A total of 4.

Note: An Electrical Installation Certificate, a Schedule of Inspection and a Schedule of Test Results must be completed for the additional ring final circuit because there is no documentation available for this circuit.

Q8. Answer: d) AC supply polarity and 1 phase 2 wire.

Reasoning: Section I: *Number and Type of Live Conductors* of the Electrical Installation Condition Report: the inspector is required to identify the number of "live" conductors, which, for a single phase supply, will be the line and neutral conductors (not the cpc) and the type of supply is AC. There is a possibility that the *confirmation of the supply polarity box* could be overlooked.

Note: the line and neutral conductors become live conductors only when the circuit's load draws current.

Q9. Answer: c) The highest measured short circuit current is greater than Icn; an Observation Code should be raised.

Reasoning: The value of the prospective fault current (Ipf) is not given in the scenario for this installation, therefore it must be calculated from the data given.

The nominal voltage for the installation is 230 volts rms and the external impedance (Ze) is 0.03Ω, using Ohms Law equation:

I = V/Ze = 230/0.03 = 7667A or 7.7kA (rounded up to one decimal place).

This value, 7.7kA, exceeds both Icn and Ics (see *GN3 Table 2.8 page 71*). Consequently there is a potential danger that merits an Observation Code of C2.

Q10. Answer: d) 6

Reasoning: An additional protective device is a 30mA RCD; therefore it will be tested on the first and second half cycle of the input supply sinewave (0° and 180°), as follows:

½I∆n 15mA ≥ 2 sec (twice)

I∆n 30mA ≤ 300 ms (twice)

5I∆n 150mA ≤ 40 ms (twice)

A total of 6 test operations.

Furthermore, a 30mA RCD is the only RCD that should be tested at five times its nominal residual current, (see Appendix 3 Table 3A page 363 Regulation 643.8 page 235).

Q11. Answer: d) The mechanical function of the device.

Reasoning: The integral test button of an RCD has only one function; that is, to test the mechanical function of the device, and will not operate unless the RCD is energised (see *GN3 paragraph 2.6.19 pages 75 & 76*).

Q12. Answer: b) 10A

Reasoning: The current version (4th Edition) of GS 38 stipulates the use of 10A fuse protection for the test instrument's probes when conducting either an RCD or earth fault loop impedance test.

Q13. Answer: d) Carryout an Initial Verification; confirm the circuit complies with the current wiring Regulations.

Reasoning: Although the ring final circuit was installed 5 years ago and has been in constant use, the inspector cannot assume that the circuit meets the requirements of BS 7671 or it is safe to use; although it is in current use. Therefore the inspector must carry out an Initial Verification and complete the required documentation: Electrical Installation Certificate and the associate Schedules.

Q14. Answer: d) Continuity of protective conductors including bonding conductors and, continuity of ring final circuit conductors, insulation resistance and SELV protection: Low resistance ohmmeter and insulation resistance tester.

Reasoning: The sequence of instrument tests is given in Part 6 Regulation 643.2 pages 232–233 and GN3 paragraph 2.6.4 page 34; the type of instruments to be used are given in GN3 Section 4 pages 109–110.

Q15. Answer: c) 16mm^2

Reasoning: The csa of the bonding conductor is given in the scenario but not that of the line conductor; however in Table 54.8 page 203 of BS 7671, there is a clear correlation between the neutral and bonding conductors. Furthermore, for a bonding conductor with a csa of 10 mm^2 the neutral will be $\leq$ 35 mm^2; therefore the line conductor must have a similar csa to that of the neutral.

Similarly, Table 54.7 on page 198 relates the csa of the line conductor to the main earth. Where the line conductor is greater than 16 mm^2 and $\leq$ 35 mm^2 the earth conductor will have a csa of 16 mm^2.

Q16. Answer: d) Take action because the device can interrupt safely without a loss of performance. Issue an Observation Code C3.

Reasoning: This type of scenario can create a slight dilemma for the inspector; although the Ics value <u>has</u> been exceeded, the Icn <u>has not</u>; therefore the characteristics of the protective device have not been compromised. Consequently, the circuit breaker will continue to provide fault protection; however the client should be notified in the form of the Observation Code C3 (see *GN3 Table 2.8 pages 71 & 72*).

Q17. Answer: d) RCD six monthly, non-standard cables and Periodic Inspection and Testing.

Reasoning: Section 4 of the Condition Report Inspection Schedule identifies <u>four items</u> that require a warning notice, but <u>only three are necessary</u> for this installation. They are:

i) **4.10 – Presence of RCD six monthly test notice** (RCBO is used for additional ring final circuit).

ii) **4.11– Presence of non-standard (mixed) cable colour notice** (the original installation was constructed 15 years ago using Red, Black and Green coloured cables; the change of cable colour was introduced in 2005 and became standard in 2006. The additional ring final circuit was constructed

5 years ago, therefore Brown, Blue and Green/Yellow coloured cables would have been used.

iii) **4.13 – Presence of other required labelling** (please specify). (Item 4.13 requires the inspector to specify; that is, <u>identify</u> any other labels. The only other notice that meets this criteria is the Periodic Inspection and Testing notice.)

iv) **Item 4.12 is <u>not</u> applicable.**

(See *BS 7671 Appendix 6 page 481.*)

Q18. Answer: d) IP2X or IPXXB and IP4X or IPXXD.

Reasoning: The question refers to the DB's enclosure, which also includes the horizontal top surface of the distribution board; therefore <u>all</u> the IP Codes given in Regulations 416.2.1 and 416.2.2 page 74 are necessary.

Q19. Answer: b) 1.29MΩ.

Reasoning: When measuring the overall insulation resistance, the individual circuit's insulation resistance form a parallel network that can be resolved using the following formulae:

$$\frac{1}{R_T} = \frac{1}{R_1} + \frac{1}{R_2} + \frac{1}{R_3} + \frac{1}{R_4} + \frac{1}{R_5} + \frac{1}{R_6} + \frac{1}{R_7} + \frac{1}{R_8} + \frac{1}{R_9}$$

$$\frac{1}{R_T} = \frac{1}{200} + \frac{1}{200} + \frac{1}{150} + \frac{1}{50} + \frac{1}{50} + \frac{1}{200} + \frac{1}{50} + \frac{1}{1.5} + \frac{1}{200}$$

R1 to R9 represent the individual insulation resistive values.

Take the reciprocal of each individual value, for example 1/200 = 0.005:

1/RT = 0.005 + 0.005 + 0.007 + 0.02 + 0.02 + 0.005 + 0.04 + 0.67 + 0.005

1/RT = 0.777

RT = 1/0.777 = 1.289 MΩ Two places of decimal 1.29MΩ

Provided the answer is less than the lowest individual insulation resistive value, then the candidate can be confident he/she has the correct answer.

Q20. Answer: d) The disconnection times meet those stated in Chapter 41 and highlight other defects.

Reasoning: The fundamental objective of inspecting and testing is to ensure an installation is in a satisfactory condition and it can continue to be used in a safe manner. To assist in this process, the disconnection times of the installation's protective devices must comply with those in given in Chapter 41of BS 7671. Failure to meet the required disconnection times can highlight other defects.

(See *GN3 paragraph 3.8 page 88.*)

Initial Verification and periodic inspection: practical

All C&G 2391-52 candidates will carry-out an Initial Verification and a Periodic Inspection on a pre-constructed installation fabricated to City & Guilds design and standard.

The candidate will be provided with a full description of the electrical installation to be inspected and tested that should be read carefully; the data within the scenario is essential and necessary to carry-out the visual inspection and testing of the installation.

The testing sequence must follow the one specified in BS7671 and GN3; both documents may be referred to during the testing procedure to determine compliance.

BS7671 Electrical Installation Certificate (EIC) and Electrical Installation Condition Report (EICR), with associated Schedules of Inspection and Schedules of Test Results documentation will be provided

Candidates may use their own test instruments; however it is their responsibility to ensure the accuracy of the instrument and compliance with the current Edition of GS 38. If the instrument, associated leads and probes fail to comply the assessor may not permit their use during the assessment.

An open book, short answer question paper, relating to the practical concepts of the assessment will be completed by the candidate usually before commencing with the practical assignment.

The open book aspect of the question paper (not multi-choice) allows reference to the following documents: BS 7671, On-Site-Guide and GN3

Typical Questions

Q1. Explain why the inspector must complete the limitations item described in Section D of an Electrical Installation Condition Report but not on an Electrical Installation Certificate where there is no equivalent section.

Response

It may not be possible to carry-out the testing procedure on given circuits within a functional installation for operational reasons of the installation. It is the responsibility of the inspector to record the condition of the installation including limitations which have been agreed with a named person.

There can be no limitations on a new installation: all circuits must be inspected and tested before an EIC can be issued

Q2. Explain why the original copy of an Electrical Installation Certificate must be retained by the client and a copy by the inspector

Response

An Electrical Installation Certificate, with related Schedules, confirms an electrical installation has been designed, constructed, inspected and tested in accordance with BS 7671.

The inspector retains a copy of the work carried-out for his/her own records to confirm, for example, the inspection and testing conforms to BS 7671

Q3. List the main steps that would be undertaken to ensure an electrical installation is securely isolated from the supply using the Main Switch of a single-phase distribution board (DB).

Response

- Inform relevant person the installation or part of it is to be isolated
- Identity the relevant DB
- Switch the supply off
- Lock the main switch off
- Confirm the circuit and the equipment dead using an approve voltage tester
- Display warning notices and erect barrier where necessary

Q4. Explain who may be in danger, and why, if the safe isolation of an installation was not undertaken before inspection and testing

Response

- The inspector: Other contractors could reinstate the supply when they come on site
- Individuals: If they are unaware testing is taking place they could be in danger during live testing

Q5. What action should an inspector take if any item of equipment he/she considers unsatisfactory during the inspection of new installation

Response

- Any defects or omissions revealed by the inspector must be rectified by the contractor
- When the defects or omissions have been rectified the installation must be inspected and retested
- The EIC must not be issued until the installation conforms to BS7671

Q6. State the Observation Code that should be given for each of the following when carrying-out a Periodic Inspection:
- a short circuit between the line and neutral conductors of a ring final circuit
- a circuit supplying radial socket outlets in a domestic dwelling with no RCD protection
- the insulation reading on a lighting circuit is less than the minimum value given in BS7671

Response

- C1 Danger present
- C2 Potentially dangerous
- F1 Further investigation required without delay

Q7. Give three areas of responsibility, which requires a signature to confirm the given area meets the *standard*, on an Electrical Installation Certificate

Response

- Design
- Construction
- Inspection & Testing

Q8. Give <u>three</u> physical items which an instrument's <u>probes</u> must have to comply with GS 38

Response

- exposed metallic probes tips which do not exceed 4mm
- finger guards
- protective fuse either 500mA or 10A depending on the testing operation
  The question requires three physical items which could include a CAT number, this must be either III or IV

Q9. State clearly the meaning of IP4X and where it applies

Response

The number 4 identifies the maximum hole/aperture of 1mm to give protection against, for example, the ingress of small pieces of wire; whereas X relates to the protection against the ingress of water in this example the level of protection has not been allocated.
The code applies to the top horizontal surfaces of enclosures

Q10. State the action the inspector should take, when during the initial verification, a final circuit is recorded as unsatisfactory

Response

The inspector should informer the installation's installer who should rectify the defect; the inspector will re-test the circuit and provided the final circuit is satisfactory, issued the Electrical Installation Certificate

8
Further questions

1. Which Statutory Regulation defines the status and title of an inspector?
   a) Health and Safety at Work etc Act 1974
   b) Electricity at Work Act 1989
   c) Electricity at Work Regulations 1989
   d) Electrical Wiring Regulations.

2. How is the competence of an inspector determined?
   a) Sound knowledge and experience
   b) Sound knowledge and experience of testing
   c) Sound knowledge and relevant experience of testing
   d) Sound knowledge, relevant experience of testing and technical standards.

3. What are the responsibilities of the Duty Holder during the Inspection and Testing process?
   a) Manage the workforce efficiently
   b) Manage associated risks safely
   c) Manage and liaise with the client
   d) Manage calibration of equipment regularly.

4. Which Statutory Regulation is applicable to cutting off the supply for safe isolation of an electrical installation?
   a) 11
   b) 12
   c) 13
   d) 14.

5. Which Statutory Regulation is referred to as the *Defence Regulation*?
   a) 15
   b) 16
   c) 21
   d) 29.

6. Which non-statutory document should the inspector refer to with regards to the test instruments used for testing an electrical installation?

a) GS3

b) GN3

c) GS38

d) GS85.

7   Which document should the inspector give to a client after a consumer unit has been replaced?
a) Minor Electrical Installation Works Certificate
b) Electrical Installation Certificate
c) Periodic Inspection Certificate
d) Electrical Installation Condition Report.

8   Which document should an inspector hand to a client after an additional lighting point has been fitted to an existing circuit?
a) Minor Electrical Installation Works Certificate
b) Electrical Installation Certificate
c) Periodic Inspection Certificate
d) Electrical Installation Condition Report.

9   A 13A BS1363 socket outlet has been included in a radial circuit without additional protection. What document(s) should be handed to the client?
a) Minor Electrical Installation Works Certificate
b) Electrical Installation Certificate
c) Minor Electrical Installation Works Certificate and a Risk Assessment Document
d) Electrical Installation Certificate and a Risk Assessment Document.

10   Additional protection is required for socket outlets with a rating not exceeding 32A. Which type of installation is not exempt from this requirement?
a) Commercial
b) Industrial
c) Domestic
d) Exhibition Shows.

11   An Electrical Installation Condition Report should be issued for which situation?
a) The addition of a socket outlet to an existing circuit.
b) A change of use of a building from a shop to a dwelling
c) The installation of a new consumer's control unit (CCU)
d) A new circuit supplying an air-conditioning unit in an office.

12   What process would provide an engineering view on the continued safe use of an installation?
   a) Initial Verification and issuing an Electrical Installation Certificate
   b) Initial Verification and issuing an Electrical Installation Condition Report
   c) Periodic Inspection and Testing and issuing an Electrical Installation Certificate
   d) Periodic Inspection and Testing and issuing an Electrical Installation Condition Report.

13   The csa and length of cables supplying a radial power circuit are doubled. What will the overall resistance of the circuit be?
   a) Doubled
   b) Halved
   c) No change
   d) A significant change.

14   Determine the total resistance, at 20°C, of a radial circuit's 2.5 mm² line conductors with a 50 m run.
   a) 0.37Ω
   b) 0.015Ω
   c) 0.741Ω
   d) 0.37mΩ

15   Determine the earth fault loop impedance for a circuit protected with a Type B 32A when measured at 30°C ambient temperature.
   a) 1.09Ω
   b) 1.37Ω
   c) 1.44Ω
   d) 1.30Ω

16   What is a suitable method of testing the function of an insulation resistance tester?
   a) On a known supply
   b) Open/closed circuit test
   c) Testing on a length of twin cable
   d) Checking the date of the last calibration.

17   A Type 3 SPD (incorporated in the socket outlet) cannot be disconnected. What action should the inspector take when conducting an insulation resistance test?
   a) Test at 500 volts dc with an insulation resistance reading of ≤ 1MΩ
   b) Test at 500 volts rms with an insulation resistance reading of ≥ 1MΩ

c) Test at 250 volts dc with an insulation resistance reading of $\geq 1\text{M}\Omega$

d) Test at 250 volts dc with an insulation resistance reading of $\leq 1\text{M}\Omega$.

18 When can the test sequence for a Periodic Inspection not follow the test sequence for Initial Verification?
   a) The installation is functioning and no faults are present
   b) Tests should be arranged to minimise disturbance
   c) Previous test data shows no faults are present
   d) Dead tests are only required to confirm compliance.

19 What action should an inspector take if an electrical circuit warrants an Observation Code Cl during a periodic inspection?
   a) Investigation work should be undertaken within ten working days
   b) Remedial work should be undertaken before any new works
   c) Remedial work should be undertaken immediately
   d) Investigation work should be undertaken urgently.

20 Electrical Installation Certificates and Minor Electrical Installation Works Certificates should be complied, signed and authenticated by:
   a) A skilled person (electrical) skilled in such work
   b) The inspector
   c) A competent skilled person skilled in such work
   d) A skilled person competent to verify the Standard.

21 A Periodic Inspection is required but there is no documentation available. What action should the inspector take?
   a) Make a note of this in the limitation section
   b) Complete a 100% survey
   c) Carry out a 100% exploratory work of the installation
   d) Carry out a degree of exploratory work including a survey.

22 What factors affect the voltage drop in a circuit?
   a) Design current and cable length
   b) Current carrying capacity of the cable and cable length
   c) Current rating of the protective device and cable length
   d) Design current, cable length and its resistance.

23 A Main Protective Bonding Conductor connection has been built in and is not accessible for testing. What is the recommended maximum value of resistance, as a benchmark, when testing between adjacent extraneous conductive parts?
   a) 0.05Ω
   b) 0.50Ω

   c) 1.00Ω

   d) 0.10Ω

24 What is the significance of the continuity reading taken at the extreme electrical point of a circuit using Test Method 1?
   a) It is the highest resistive reading
   b) It confirms continuity
   c) It is the reading recorded on the Schedule of Test Results
   d) It is used to confirm the voltage drop.

25 A durable copy of the installation's schedule is provided within the system's main DB, which indicates an item of equipment is vulnerable to the insulation test voltage and cannot be isolated. What action should the inspector take?
   a) Test at a reduced voltage
   b) Test with the neutral and cpc connected together with the line conductor
   c) Test with the neutral and line connected together with the cpc
   d) Test with the line and neutral connected together with the protective Earth.

26 Determine the overall insulation resistance where the individual readings for four circuits are: 200MΩ, 100MΩ, 50MΩ and 20MΩ.
   a) 1.68MΩ
   b) 16.8MΩ
   c) 16.8Ω
   d) 168Ω

27 An inspector should carry out an inspection of an installation's electrical equipment before conducting the test sequence. Which of the following is not part of this inspection process?
   a) They are compliant with Section 511
   b) They are correctly selected and erected
   c) They are partially dismantled
   d) They are not damaged or defective.

28 In a location with a protective bonding system to Regulation 411.3.1.2, containing a bath, supplementary bonding may be omitted, provided:
   a) All pipe work, in the location, is a mixture of pvc and copper
   b) All extraneous-conductive parts meet the Regulation 411.3.1.2
   c) All final circuits of the location meet the requirements of Regulations 411.3.1.2 and 701.411.3.3
   d) All final circuits and extraneous-conductive parts meet the appropriate Regulations for safety in the location.

29 What is the test voltage and minimum acceptable resistance for SELV circuits when conducting an insulation resistance tested between line and earth conductors?
   a) 500 volts dc and 1MΩ
   b) 250 volts dc and 1MΩ
   c) 500 volts dc and 0.5MΩ
   d) 250 volts dc and 0.5MΩ.

30 What data should be recorded for a circuit's protective device?
   a) In, type and breaking capacity
   b) Ics, type, breaking capacity and BS Number
   c) Icn type, breaking capacity and BS Number
   d) In, type, breaking capacity and BS Number.

ANSWERS SOURCE

1	c)	EWR Memorandum pages 11& 39 Regulations 3 & 16
2	d)	GN3 para. 1.2 page 12
3	b)	EWR Memorandum page11 Regulation 3
4	b)	EWR Memorandum page 31
5	d)	EWR Memorandum page 41
6	c)	Electrical Test Equipment 4th Edition
7	b)	BS 7671 page 462 & 463
8	a)	BS 7671 page 465
9	c)	BS 7671 pages 59 & 465
10	c)	BS 7671 Reg. 411.3.3 page 59
11	b)	GN3 pages 81 & 82
12	d)	GN3 para.3.1 page 81
13	c)	Calculation $R = l/A = 2 \times l/ 2 \times A = l/A$ where A = csa
14	a)	GN3 page 43 $7.41/100 \times 50 = 0.37\Omega$
15	b)	BS 7671 Table 41.3 page 62
16	b)	Practical: standard test
17	c)	BS 7671 Part 6 page 232
18	c)	GN3 para. 3.10.1 page 98
19	c)	BS 7671 Item 7 page 476
20	d)	BS 7671 Reg 644.5 page 236
21	d)	GN3 para. 3.8.3 pages 89 & 90
22	d)	BS 7671 Appendix 4 Item 6 page 381 & 382
23	a)	GN3 page 45
24	c)	BS 7671 page 483 and GN3 page 42

25	c)	GN3 page 52
26	b)	Calculation
27	c)	BS 7671 Part 6 Reg.642.2 page 230
28	d)	BS 7671 Section 701 Reg. 701.415.2 page 241
29	d)	GN3 Table 2.4 page 54 and BS7671 Table 64 page 232
30	d)	BS 7671 page 483 (In ≡ rating (A))

9
Equations and calculations

INTRODUCTION

Solving an arithmetical problem should not be a daunting experience. After all, there are only four functions that can be applied to any arithmetical operation. They are: addition, subtraction, division and multiplication; whereas writing a detailed answer to a question requires the manipulation of 26 letters!

Probably the foremost difficulty a candidate may experience, when faced with a question that requires an element of calculation, could be, initially, remembering which particular formula to use to resolve a problem. If this is an obstacle, reverting to rote could overcome it: whereby formulae are constantly repeated, ideally in a written format, until they are firmly embedded in a candidate's memory cells.

Another possible problem could be the re-arranging, or transposition, of an equation, which can be explained as follows.

Take the simple equation 2 + 4 = 6. The equal sign (=) merely states that the left-hand side of the equation is equal to the right-hand side. Consequently, if 2 is subtracted from 6, on the right-hand side of the equation, will it remain equal?

Clearly, subtracting 2 from the right-hand side (2 + 4 = 6 − 2) will not result in equality (2 + 4 ≠ 4) simply because the left-hand side of the equation is not equal to the right-hand side. Therefore, to achieve equality, the arithmetical operation which has taken place on the right-hand side of the equation must also be applied to the left-hand side, that is: subtract 2 from both sides of the equation, as follows: 2 − 2 + 4 = 6 − 2, which will result in equality: 4 = 4.

Consequently, the arithmetical operation which needs to take place on one side of the equation must also be applied to the other side. Also, do not be phased by letters in place of actual numbers; letters are there to represent a quantity, for example the earth fault loop impedance is a relatively common electrical equation where: $Zs = Ze + (R_1 + R_2)$, which, hopefully, should not be a problem.

Note: A number or a letter is considered to be positive unless otherwise stated. For example, Zs and Ze are both positive quantities but without the positive symbol (+). A simple case of sign character economics.

Example: Determine the value of $(R_1 + R_2)$ in terms of Zs and Ze:

- Subtract Ze from both sides of the equation: $Zs - Ze = Ze - Ze + (R_1 + R_2)$
- $Zs - Ze = (R_1 + R_2)$ or $(R_1 + R_2) = Zs - Ze$.

The equation can be re-arranged either way because both sides are equal.

SUBSTITUTION

If $Ze = 0.8\Omega$ simply replace Ze in the equation $Zs = Ze + (R_1 + R_2)$ with its numerical value.

- Where $Zs = 0.8 + (R_1 + R_2)$

Subtract 0.8 from both sides of the equation:

- $Zs - 0.8 = 0.8 - 0.8 + (R_1 + R_2)$, which effectively eliminates 0.8 from the right-hand side:

Then $R_1 + R_2 = Zs - 0.8$.

Hopefully, this chapter will assist, and encourage, those candidates who are conscious of their modest or poor arithmetical abilities.

9.1 Cable resistance

Every metallic conductor will have a resistive value per metre (mΩ/m), which will depend on its cross sectional area, whereas the overall resistance of a conductor will depend upon other factors such as:

- length
- cross sectional area
- temperature.

Probably temperature, both ambient and conductor temperatures, can be regarded as a significant contributing factor; with 20°C being the optimum value for calculations and testing purposes unless otherwise stated.

(See *Regulation 132.6(i): Cross Sectional Area of Conductor BS 7671 Chapter 13.*)

Accordingly, IET's *On-Site-Guide Appendix I* illustrates the ohmic values of line (R_1) and protective conductors (R_2) at the optimum value of 20°C. Consequently, this temperature level is generally used as the benchmark for determining the overall resistance of final circuits' conductors; whereas the maximum tabulated values given in BS 7671 are determined at the 30°C level. (Also see *GN3 Appendix A Earth Fault Loop Impedance*.)

Conductor resistivity

The resistance (R) of a metallic conductor is equal to its length (L) and inversely proportional to its cross section area (csa). Consequently if the length of the conductor is increased, its resistance will increase; if the cross sectional area is increased, the conductor's resistance will decrease where the temperature (T) remains constant.

Therefore R = L/csa

Example 1

Question: The resistance of a length of cable is 1.0Ω. What would its resistance be if its length and its cross sectional area were doubled?

Result: The resistance would remain the same. Doubling the cable length = 2 × 1.0 = 2 and doubling the cross sectional area would effectively halve the total resistance 2/2 = 1.0Ω.

Example 2

Question: Determine the resistance of a radial circuit's protective conductor (R_2). The length of the 2.5 mm² diameter conductor is 50 metres and the resistance, at 20°C, is 7.42mΩ/m; that is, 7.42 milliohms for every metre of conductor.

7.42mΩ is equal to $7.42 \times 10^{-3}\Omega$ which can be re-written as 7.42/1000.

If the candidate inputs 1/1000 into an electronic calculator, the result will be either 10^{-3} or 0.001; therefore $10^{-3} = 1/1000$.

The total resistance of the conductor will be its length (L) multiplied by its resistance per metre (mΩ/m), which is equal to: 50 × 7.42/1000 = 0.371Ω.

Example 3

Question: A ring final circuit is installed using 2.5/1.5 twin + earth cable, protected with a RCBO 30mA/32A BS EN 61009, with a total cable length of 60 metres. Determine the *expected* value of $(R_1 + R_2)$ at each socket outlet.

The resistance per metre, at 20°C, for the line conductor and its cpc is 7.42mΩ and 12.10mΩ respectively.

This Example is slightly different from Example 1 inasmuch as the candidate is required to determine the *expected* value of both the line and protective conductors $(R_1 + R_2)$, at each socket outlet on a ring final circuit. Because the value of $(R_1 + R_2)$ is determined by calculation, the resulting value can only be an *expected* value, whereas the measured value could be slightly different depending on the conductor's characteristics and ohmic connections.

If the final circuit had been a radial type, the arithmetical process would have been relatively simple, inasmuch:

$$*(R_1 + R_2) \times L = \frac{(7.42 + 12.10)}{1000} \times 60 = \frac{19.52 \times 60}{1000} = 1.17\Omega*$$

Note: 7.42mΩ can be expressed as either 0.00742Ω or 7.42Ω/1000 the latter tends to be used more frequently

But this question relates to a ring final circuit, and there are two slightly different methods that may be applied to resolve the problem. They are:

Method 1, which is effectively the reverse of Steps 1 and 3 of *Test 2: Continuity of a Ring Final Circuit*; whereby the end-to-end continuity testing of the line conductor (r_1) and the circuit protective conductor (r_2) are measured and recorded.

SOLUTION

The calculated ohmic value for each conductor is:

Line conductor: r_1 = length × mΩ/m = 60 × 7.42/1000 = 0.445Ω

Protective conductor: r_2 = length × mΩ/m = 60 × 12.10/1000 = 0.726Ω

Where $\dfrac{r_1 + r_2}{4}$ = $R_1 + R_2$ = $\dfrac{0.445 + 0.726}{4}$ = 0.293Ω

Method 2 is similar to Method 1, as follows:

$$R_1 + R_2 = \frac{\text{Cable length (L)} \times \text{the total resistance of the individual conductor's m}\Omega/\text{m } (r_1 + r_2)}{4}$$

$$R_1 + R_2 = \frac{r_1 + r_2}{4} = \frac{60 \times (7.42 + 12.10) \times 10^{-3}}{4}$$

For simplification purposes, the equation will be resolved in two stages:

Stage 1: resolve the quantities in the brackets.

$$(r_1 + r_2) \times 10^{-3} = \frac{(7.42 + 12.10)}{1000} = \frac{19.52}{1000} = 0.0195\Omega \text{ the value of } (r_1 + r_2)$$

Stage 2

$$\frac{(r_1 + r_2)}{4} \times L = \frac{0.0195 \times 60}{4} = 0.293\Omega$$

Note the arithmetical similarities between Method 2 and the method used to calculate the resistive value of $(*(R_1 + R_2) \times L *)$ for a radial circuit.

9.2 Voltage drop

The voltage drop for any functional final circuit cannot be measured; therefore it must be calculated using either the installation's design criteria or the final circuit's measured impedance.

Probably the most suitable equation to use is:

$$\text{Volts Drop} = (\text{mV/A/m}) \times I_b \times L$$

where the ambient temperature is given as 20° C

Where (mV/A/m) is a constant for a given type of conductor which can be found in *Appendix 4 of BS 7671*, measured in millivolts.

I_b is the circuit's design current or load current, in amperes (A).

L is the circuit's cable length, in metres (m).

Note: If the ambient temperature is given as 30° C then the correction factor 1.2 must be included in the equation:

$$\text{Volts Drop} = (\text{mV/A/m}) \times I_b \times L \times 1.2$$

(see IET On-Site-Guide Appendix I Table 13 page 198)

If the constant (mV/A/m) and the length (L) values are not available, but the circuit's impedance (R) and load current (I_L) are known, then the alternative equation to use is the Ohm's Law Equation.

Where $$V = I \times R.$$

Where V is the volts drop.

I is the load current in amperes.

R is the circuit's conductor's resistance, which includes both the line and neutral conductors.

Power Laws may also be derived from the Ohm's Law Equation, if required.

The fundamental Power Law is: Power (P) = Load Current (I_L) multiplied by the Nominal Voltage (V).

Where $$P = I \times V.$$

Example 1

Question: Determine the volts drop for a circuit where the load current is <u>20 A</u> and the resistance of the line and neutral conductors is <u>0.46Ω</u>.

Result: Use Ohm's Law Equation: $V = I \times R$ because the only information given is current (I) and circuit resistance (R), therefore:

$$V = 20 \times 0.46 = \textbf{9.2 volts.}$$

Do *not* forget to include the correct units in the final answer.

A supplementary question to this Example could be: Is the volt drop acceptable?

Initially this could be difficult because the type of circuit has not been specified; however the load current is 20 A, therefore it can be assumed that the circuit is *not* a lighting type.

The maximum volt drop for *other uses* is 5% of nominal voltage; that is:

$$5\% \text{ of } 230 = \textbf{11.5 volts} \text{ or } 5/100 \times 230 = \textbf{11.5 volts.}$$

The answer to the supplementary question would simply be a Yes, because the calculated value of 9.2 volts is below the ceiling level of 11.5 volts.

Note: The threshold voltage drop for lighting circuits is 3% of the nominal voltage, which is 6.9 volts. Furthermore, the 3% and 5% ceiling levels relate only to the public energy supply; whereas for a private supply it is 6% and 8% respectively.

(See *Appendix 4 of BS 7671.*)

Example 2

Question: The voltage drop for a 2.5 mm² multicore cable is 18 mV/A/m, if the cable length is 50 metres and the load is 2.3kW. Determine the volt drop for the single phase final circuit, assuming the ambient temperature is 20°C and the supply voltage is 230 volts rms, and state whether the volt drop is acceptable.

Note: 2.3kW = 2300 watts.

Result: Initially, use the Power Law Equation to determine the load current where

$$P = I \times V$$

Re-arrange the equation to calculate the load current (I_L). Remember the arithmetical operation carried out on the right-hand side of equation must also be carried out on the left-hand side.

Therefore, to determine the current (I), divide both sides of the equation with the voltage (V):

$$P/V = I \times V/V.$$

The voltage component will cancel on the right-hand side leaving:

$$I = P/V = 2300/230 = 10A.$$

Apply the volt drop equation: Volt Drop = (mV/A/m) $\times I_b \times L$.

Where (mV/A/m) = 18×10^{-3} volts, $I_L \equiv I_b = 10A$ and L = 50m.

$$\frac{18}{1000} \times 10 \times 50 = \textbf{9 volts}$$

On the surface, a 9 volt drop is well within the 5% of the nominal voltage ceiling (5% of 230 = 11.5 volts), however caution must be observed; does the cable length include both the <u>*Line*</u> and <u>*Neutral*</u> conductors? If *not*, then there will be a 9 volt drop on *both* the line and neutral conductors. Consequently, the net volt drop would be 18 volts, *not* 9 volts.

When answering a question on volts drop, read it carefully. If it does not refer to a combined conductor resistance then be cautious.

9.3 Earth fault loop impedance

There are a number of equations which can be used to determine the value of the *earth fault loop impedance*:

$$Zs = Ze + (R_1 + R_2) \times L \times 1.2$$

Where Ze is the *external* earth fault loop impedance.

$R_1 + R_2$ is the line and cpc resistive values.

L is the final circuit's cable length.

1.2 is a temperature correction factor for 30°C.

$$Zs = \frac{Uo \times Cmin}{Ia} \text{ and } Zs \times Ia = Uo \times Cmin$$

Where Uo is the nominal voltage.

Cmin is a correction factor to accommodate voltage variations.

Ia is effectively a fault current, which will cause a protective device to operate

$$Zs(m) = \frac{Uo \times Cmin}{Ia} \times 0.8$$

Where Zs(m) is the *measured* earth fault loop impedance.

0.8 is a temperature correction factor.

$$Ze = Uo/Ipf$$

Where Uo is the single phase nominal voltage and Ipf is the highest measured prospective fault current.

Example 1

Question: A measured value of 2.2Ω was recorded for the earth fault loop impedance (Zs) of a radial circuit protected with a 16A Type B circuit breaker to BS EN 60898. The maximum tabulated value given in BS 7671 is 2.73Ω. Is the measured value acceptable?

Result: The question gives large amount of information to select from, but most of it is superfluous. There are only two pieces of data which are of any significance: 2.73Ω and 2.2Ω.

At this juncture it is the candidate's knowledge and understanding which come into play to recognise that the 2.73Ω is the maximum tabulated value at an ambient temperature of 30°C; whereas the 2.2Ω is the measured value at ambient temperatures considered to be at 20°C.

Therefore is the measured value acceptable considering the temperature difference?

To give a definitive answer to this question will require a small calculation incorporating the temperature correction factor of 0.8; whereby the maximum tabulated value is multiplied by the temperature correction factor:

$2.73 \times 0.8 = \mathbf{2.18\Omega}$ which is the *maximum* acceptable *measured* value.

The answer could create a slight dilemma for the candidate, because the calculated value is so close to the measured value. Nevertheless the *maximum measured* value is **2.18Ω**; therefore 2.2Ω is *not* acceptable and further circuit investigations needs to be made.

It is worthwhile noting, if the candidate was on-site there is a high probability his/her test instrument would *not* display a value of **2.18Ω**; it would more than likely display 2.2Ω, depending on the instrument's resolution. Accordingly, in practice the circuit's parameters would need further investigation.

(see *BS 7671 Appendix 3 pages 362 & 363*).

9.4 Insulation resistance

When the insulation resistance of the whole consumer unit is tested, and a low result is recorded, the standard practice is to determine which circuit is causing the problem.

The practical method is to remove or open each protective device, successively, and re-test on each occasion until the resistive value returns to an acceptable level.

The question is: why should one circuit's insulation resistance have such a catastrophic effect on the remaining circuits' insulation resistance?

The reason is: when conducting the insulation resistance test, the conductor's insulation will be subjected to electrical pressure from the test instrument, which will also occur during the installation's normal operations. Either action will create a series of parallel resistive pathways and if just one pathway has a low resistance, possibly caused by insulation deterioration, the overall insulation resistance could be *pulled-down*; that is: significantly reduced.

This explanation can be illustrated in the following example; however it will require a basic understanding of parallel resistive networks and Ohm's Law.

Example 1

Question: The insulation resistance, for each final circuit, is tested between the live and circuit protective conductors. The following test results, measured in megaohms (MΩ) were recorded: 200, 200, 150, 50, 25, 100 and 2. Determine the overall insulation resistive value.

Because *all* the resistive values are given in megaohms (MΩ) simply use the values given, but include the unit in the final answer; this action will help simplify the calculation process.

While the majority of the recorded values are very high, it is the comparatively low resistive value of the 2MΩ pathway which will have the most significant effect on the system's overall insulation resistance reading; inasmuch as it will reduce or *pull-down* the overall resistive value. Therefore before conducting any arithmetic process the candidate should simply review the insulation resistive values given, then anticipate or *estimate* the overall value; which will always be equal to or less than the lowest resistive value given. In this particular example the overall calculated value should be around 2MΩ or less and, provided the calculated value is similar to that of the estimated value, the candidate can be confident that his or her answer will be correct.

Note: A candidate may be required to give an *estimated value* to an insulation resistance problem *before* conducting the arithmetic process. Furthermore, if the question states: *all calculations are to be shown* then the following example should be followed.

Resistors in Parallel

$$\frac{1}{Rt} = \frac{1}{R_1} + \frac{1}{R_2} + \frac{1}{R_3} + \frac{1}{R_4} + \frac{1}{R_5} + \frac{1}{R_6} + \frac{1}{R_7}$$

$$\frac{1}{Rt} = \frac{1}{200} + \frac{1}{200} + \frac{1}{150} + \frac{1}{50} + \frac{1}{25} + \frac{1}{100} + \frac{1}{2}$$

$$\frac{1}{Rt} = 0.005 + 0.005 + 0.007 + 0.02 + 0.04 + 0.01 + 0.5$$

Take the reciprocal of each recorded value; that is, divide the numerator by the denominator, for example: 1 divided by 200 = 0.005.

Then add all the reciprocal values, which will result in a value of 0.587.

Where 1/Rt = 0.587 to evaluate the overall resistive value (Rt) invert each side.

Rt = 1/0.587 = **1.7MΩ** which, as predicted, is around 2MΩ or less.

BS 7671 *does allows* an insulation resistance value of **1.0MΩ** as acceptable; however if the value calculated was recorded during *Initial Verification* the inspector should immediately suspect there was a problem.

Reasoning

If a further circuit, with a low insulation resistance of 2MΩ, was included in the original seven circuits tested, the overall insulation resistance would now be:

$1/Rt = 1/R_7 + 1/R_8 = 1/1.7 + 1/2$	Where $1/R_7$ (1.7MΩ) is the overall insulation resistance for the original seven circuits and $1/R_8$ (2MΩ) is the additional circuit.
$1/Rt = 0.587 + 0.5 = 1.087$	0.587 and 0.5 are the reciprocal values.
$1/Rt = 1.087$	Invert both sides of the equation.

Therefore Rt = 1/1.087 = 0.92MΩ.

Consequently, the inclusion of the low value insulation resistance circuit will have a significant effect on the whole insulation resistance; however if the circuit had a high value of insulation resistance it will have little effect.

If, for example, the insulation resistance of circuit R_8 = 200MΩ, which has a reciprocal value of 0.005, and if this value is added to the reciprocal result of the original seven circuits, then:

$$1/Rt = 0.587 + 0.005 = 0.592 \text{ and}$$

$$Rt = 1/0.592 = 1.689 MΩ \text{ which is approximately equal to } 1.7 MΩ.$$

Accordingly, if there is an increase in the number of *healthy* circuits, that is greater than 200MΩ, the overall effect on the installation's insulation resistance will be negligible; but if circuits with low insulation resistivity are included, the overall insulation resistance will be dramatically reduced.

9.5 Insulation resistivity

A conductor's insulation will *decreases* as the cable *length increases*, which is directly opposite to the resistivity of a metallic conductor. This *decrease* in insulation resistivity, as the cable *length increases*, is a direct result of *parallel resistive pathways* created when electrical pressure is applied against the insulation of the conductors under test or during normal operational conditions.

9.6 RCD used fault protection

Where an RCD is used for fault protection, two conditions should be fulfilled. They are:

- the disconnection time, as required by *BS 7671 Regulation 411.3.2.2 or 411.3.2.3 or 411.3.2.4*, shall be met
- $R_A \times I\Delta n \leq 50$ V and $R_A \leq 50/ I\Delta n$ Ω

Where R_A is the sum of the resistances of the earth electrode and the protective conductor connecting it to the exposed-conductive-parts (in ohms), and $I\Delta n$ is the rated residual operating current of the RCD (mA).

If R_A is not known, it may be replaced with Zs; and if Zs exceeds 200Ω it may not be stable.

The conditions for RCD fault protection are said to have been met if the earth fault loop impedance, of the protected circuit, meets the requirements of *BS 7671 Table 41.5*.

Problem

Determine the earth fault loop impedance of a circuit protected with a 30mA RCD. Show all calculations.

$$R_A \leq 50/ I\Delta n \ \Omega$$

$$\text{Where } I\Delta n = 30\text{mA} = 30 \times 10^{-3} = 0.03\text{A}$$

$$R_A = 50/0.03 = 1666.6\Omega \text{ or rounded up } 1667\Omega$$

SUMMARY

Arithmetical operations should not be difficult, provided the basic rules of equality are followed: the arithmetical function which takes place on one side of an equation must also take place on the other side; and remember: the opposite to positive (+) is negative (−) and vice versa.

Similarly, the opposite to multiplication (×) is division (\\) and vice versa.

Also, the quantities inside a bracket must be calculated first, followed by multiplication, then division, addition and finally subtraction. This sequence of arithmetical operations will only apply if and when required.

Problem

A 25 mm² five core, **80 m** swa cable is to be used to supply a sub-distribution board from a DB at the origin of the installation, with one of the conductors being used as the cpc. The resistance per metre is **0.727 mΩ/m** at 20°C and the external earth fault loop impedance (**Ze**) is **0.35Ω**. Determine the expected measured value of Zs. All calculations must be shown.

All relevant quantities are in **bold**.

$$Zs = Ze + (R_1 = R_2) \times L$$

The temperature correction factor 1.2 is *not* required in this calculation because the resistive value per metre (mΩ/m) is given at 20°C; it will, however, be necessary if a temperature of 30°C is quoted.

$$Zs = 0.35 + (0.727 + 0.727)/1000 \times 80$$

The ohmic value of R_2 will be the same as R_1 because *all* the cores in the swa cable will have the same cross sectional area.

$$Zs = 0.35 + (1.454/1000) \times 80$$

Determine the value of the brackets first: $(1.454/1000) = 0.001454$.

Multiply the bracket's value by the length of cable: $0.001454 \times 80 = 0.116$

$$Zs = 0.35 + 0.116 = 0.466\Omega$$

Rounded up to the second decimal place: $Zs = 0.47\Omega$

Remember, the ohmic value for Ze represents the *external* earth fault path, which is determined by its length and cross sectional area of the conductors used; whereas $(R_1 + R_2) \times L$ is effectively the *internal* fault path and Zs is the *total* earth fault loop impedance.

Further related topic example

Give three reasons for a circuit's voltage drop and the relationship.

- **<u>Length</u> of final circuit's conductors**: both the line and neutral conductor. If only the line conductor's length is given the total length will be twice that of the line (2 × line length).
- **<u>Cross sectional area of conductors</u> (csa)**: the resistance (R) of a conductor is proportional to its length (l) and inversely to its cross section area (A) [R = l/A].

The voltage dropped, in millivolt (mV), across each metre of cable will depend upon its resistivity, which in turn will depend upon its length and csa.

- **The <u>design current</u> (Ib)**: this is the current draw by the circuit's load.

Characteristics Relationship: Volts Drop = (mV/A/m) × I_b × L

Where (mV/A/m) is millivolt drop per ampere per metre, in other words there will be a small voltage drop for every ampere of current passing through one metre of cable.

The small voltage drop will depend on the csa of the conductor, the current drawn by the final circuit's load (I_b) and the length of the final circuit's conductors (L).

All three items are linked as follows:

- If the length of the circuit's conductors are increased, there will be an increase in the overall resistance.
- Therefore, there will be an increase in the total voltage drop.
- This is because of the small voltage drop for every metre of conductor for every ampere of current flowing.
- Consequently, if the load current increases the voltage drop will also increase.
- If the csa of the conductor is doubled, for example, and the load current and cable length remain constant the voltage drop will be halved.

There are a number of other factors which could affect voltage drop in an installation but this approach is possibly the simplest.

A reasonable understanding of Ohm's Law Equation would also assist a candidate, where V = I × R, and if either the current flowing through a circuit or its resistance is altered, or both, then the circuit voltage will also change.

10
C&G 2391-50,
51 and 52 syllabi

INTRODUCTION

City & Guilds publishes separate Unit syllabi for their 2391-50: *Initial Verification*, 2391-51: *Periodic Inspection* and 2391-52: *Initial and Periodic Inspection and Testing*. Although there are numerous similarities between these three Units, their divergence will be highlighted in the following summary.

Each Unit's syllabus has been designed to assist a candidate to enhance their understanding of the principles, practices and legislation, embracing the requirements of statutory and non-statutory regulations, for either or both *Initial Verification* and *Periodic Inspection* of electrical installations.

The aim of this summary is to underline the knowledge required by *City & Guilds* and to reinforce a candidate's skills for the inspection, testing, commissioning and certification of electrical installations.

SUMMARY
C&G 2391-50: *Initial verification*

The candidate will need to:

i) Understand the **requirements** for *Initial Verification* of electrical installations. To achieve this objective the candidate must be able to:
   - identify the situations that require initial verification
   - understand the purpose of initial verification
   - identify the statutory and non-statutory documents that affect the initial verification process
   - identify the information required by an inspector in order to carry out initial verification in accordance with BS 7671 and GN3
   - describe the purpose of certificating documents and the requirements for recording and retention in accordance with BS 7671
   - define the responsibilities of the relevant signatories in relation to certification.

ii) Understand **safety management** procedures when undertaking initial verification:
  - identify health and safety requirements which apply when carrying out initial verification
  - outline the relevant requirements of the Electricity at Work Regulations 1989 (EWR) for safe Inspection and Testing of electrical installation
  - describe the procedure for completing safe isolation in accordance with HSE guidance
  - explain why safe isolation is carried out for the protection of the inspector and other persons
  - identify the implications of not carrying out safe isolation.

iii) Understand the **requirements** for the initial inspection of an electrical installation:
  - state why inspection is carried out before testing
  - identify the items to be inspected during initial verification in accordance with the Schedule of Inspections for new installations
  - explain how the items on the Schedule of Inspections would be verified
  - describe the appropriate human sense required for a particular inspection
  - describe how to deal with an item that is found unsatisfactory during the inspection process
  - describe how the IP classification system is used for electrical equipment.

iv) Understand the **requirements for testing** electrical installations at initial verification:
  - state the reasons for instruments to be maintained and comply with standards such as BS 7671 and GS38
  - describe the characteristics of instruments and leads used for each test
  - explain why there is a recommended test sequence for initial verification
  - explain the purpose of each test
  - describe how each test is carried out
  - identify factors that affect the result of each test
  - interpret the result of each test
  - explain how test results are verified for compliance.

C&G 2391-51: *Periodic inspection*

The candidate will need to:

i) Understand the **requirements** for Periodic Inspection and Testing:
- state why Periodic Inspection and Testing may be required
- state the purpose of Periodic Inspection
- identify the statutory and non-statutory documents that affect the Periodic Inspection process
- identify information to be provided in order for an inspector to agree with the client:
  a) extent and limitations
  b) appropriate sampling
- describe the purpose of the Periodic Inspection documentation and the requirements for recording and retention
- define the responsibilities of the inspector in relation to the report.

ii) Understand **safety management procedures** when undertaking Periodic Inspection and Testing:
- identify health and safety requirements that apply when carrying out Periodic Inspection and Testing
- outline the relevant requirements of the Electricity at Work Regulations (EWR) for safe Inspection and Testing of electrical installation
- describe the procedure for completing safe isolation in accordance with HSE guidance
- explain why safe isolation is carried out for the protection of the inspector and other persons
- identify the implications of not carrying out safe isolation.

iii) Understand the **requirements** for the Periodic Inspection of an electrical installation:
- identify the items to be inspected during Periodic Inspection in accordance with the Schedule of Inspections
- explain how the items on the Schedule of Inspections would be classified
- describe the appropriate human senses required for a particular inspection
- describe how to classify an item that is found to be unsatisfactory during the inspection process
- describe how the IP classification system is used for electrical equipment.

C&G 2391-52: *Initial verification and periodic inspection and testing*

The candidate will need to:

i) Understand the **requirements for Inspection and Testing**:
- state the purpose of the following:
  a) Initial Verification
  b) Periodic Inspection and Testing
- explain reasons for conducting types of Inspection and Testing
- compare initial verification and Periodic Inspection and Testing processes
- identify statutory and non-statutory documents that may be required during the Inspection and Testing processes
- identify documents that would be completed and issued following:
  a) Initial Verification
  b) Periodic Inspection and Testing
- identify information needed in order to:
  a) carry out initial verification in accordance with BS 7671/GN3
  b) agree extent and limitations
  c) determine appropriate sampling
- define the responsibilities of the relevant signatories in relation to certification and reporting
- identify the type of information to be recorded on documents for the following:
  a) Initial Verification
  b) Periodic Inspection and Testing
- describe the purpose of the Inspection and Testing documents and the requirements for recording and retention.

ii) Understand the **safety management procedures when undertaking** Inspection and Testing:
- identify health and safety requirements that apply when carrying out inspection and testing
- outline the relevant requirements of the Electricity at Work Regulations (EWR) for safe Inspection and Testing of electrical installations
- describe the procedure for completing safe isolation in accordance with HSE guidance
- explain why safe isolation is carried out for the protection of the inspector and other persons
- identify the implications of not carrying out safe isolation.

iii) Understand the **requirements for testing** electrical installations:

- state the reasons for instruments to be maintained and comply with standards such as BS 7671 and GS38
- describe the characteristics of instruments and leads used for each test
- explain why there is:
  a) a recommended test sequence for initial verification
  b) not a recommended test sequence for Periodic Inspection and test
- explain why certain tests may not be necessary at Periodic Inspection and test
- explain the purpose of each test
- describe how each test is carried out and how it may be adapted for periodic conditions
- identify factors that affect the result of each test
- interpret the result of each test
- explain how test results are verified for compliance or classification.

Glossary of terms

barrier: A part providing a degree of protection against inadvertent contact with a live part.

basic insulation: An insulating material applied to a live part(s) to provide basic protection.

basic protection: A means of protection against electric shock under fault free conditions.

bonding conductor: A protective conductor providing equipotential bonding; where:

(i) **equipotential bonding** is a means of maintaining exposed-conductive-parts and extraneous-conductive-parts at the same potential.

(ii) **exposed-conductive-parts** are the metallic casings of electrical equipment which, under fault free conduction can be handled freely but under fault conditions can become live unless adequate safety precautions are not taken to prevent this condition. In other words the user could be *exposed* to an electric shock under fault conditions.

(iii) **extraneous-conductive-parts** are those metallic parts, such as water pipes, that are liable to introduce a potentially general earth but they do not form part of an electrical installation.

(iv) **supplementary bonding** is considered to be an addition to fault protection where all simultaneous, accessible *exposed-conductive-parts* of fixed equipment and *extraneous-conductive-parts* are at the same potential. Typically, each length of metal In trunking is bonded, linked with a bonding conductor, to the system's MET. (Note: exposed structural metalwork and, where practical, any reinforcement metalwork in constructional reinforced concrete are also bonded to the system's MET.)

circuit breaker: A device constructed to carry and isolate load currents, with the capacity to automatically open (break) under predetermined conditions; they are: Icn which is the maximum fault current a device can interrupt safely but may have to be taken out of service; whereas Ics is the maximum fault current a device can interrupt safely and still remain in service.

circuit protective conductor (cpc): A protective conductor that connects exposed-conductive-parts to an installation's MET.

Class I equipment: Needs to be earthed (basic protection).

Class II equipment: Does not need to be earthed (supplementary insulation is provided).

consumer unit (CU): Generally associated with domestic premises.

current-carrying capacity of a conductor: This is the maximum current a conductor can carry under defined conditions e.g. where its steady-state temperature level is not being exceeded.

distribution board (DB): Generally associated with commercial and industrial installations.

distribution circuit: A circuit supplying a distribution board or switchgear.

duty holder: The status given in *law* to an inspector.

earth: This is regarded as the conductive mass of global earth, which is conventionally taken as zero potential.

earth electrode: A conductive part, a metallic rod, which generally is embedded in the soil, but may be other conductive medium in electrical contact with the earth embedded in, for example concrete.

earth fault loop impedance: The impedance of a loop which carries a fault current starting and ending at the point of earth fault, which is denoted by the symbol Zs.

earthing: The means of connecting the exposed-conductive-parts of an installation to its MET.

earthing conductor: A protective conductor that connects an installation's MET to either an earth electrode or a DNO's earthing network.

enclosure: This is an accessory that provides all round basic protection against external influences.

inspection: An examination of an electrical installation applying one or more of the human senses - hearing, touch, smell or sight as appropriate.

isolation: A function intended to cut off for reasons of safety the supply from all, or a discrete section, of the installation by separating the installation or section from every source of electrical energy.

line conductor: This is regarded as a conductor in an a.c. transmission system but does not include a neutral conductor, a protective conductor or a PEN conductor. When applied to the consumer's side of an installation is not an energised conductor.

live part: Whereas a live conductor is one which is energised during normal usage, including its associated neutral conductor but not a PEN conductor.

prospective fault current (Ipf): The possible fault current that can arise from a short circuit between either live conductors or live conductor's to earth.

protective conductor (PE): A conductor used for some measures of protection against electric shock and intended for connecting together any or all of the following parts:

(i)   Exposed-conductive-parts;
(ii) Extraneous-conductive-parts;
(iii) The main earthing terminal;
(iv) Earth electrode(s);
(v)   The earthed point of the source, or an artificial neutral.

skilled person (electrically): Person who possesses, as appropriate to the nature of the electrical work to be undertaken, adequate education, training and practical skills, and who is able to perceive risks and avoid hazards which electricity can create. The legal status given to an inspector is a *person competent to do such work.*

system: An electrical system consisting of a single source or multiple sources running in parallel of electrical energy and an installation.

TN-C-S system: A system in which the distributor's neutral and protective functions are combined in a single conductor but separated on the consumer's part of the system with an external earth fault loop impedance $Z_e \leq 0.35\Omega$

TN-S system: A system having separate neutral and protective conductors throughout the system with an external earth fault loop impedance $Z_e \leq 0.8\Omega$

TT system: A system that uses global earth as the fault current path where $R_A < 21(1$ or earth fault loop impedance $Z_s \leq 200\Omega$ If this value is exceeded it is regarded as being unstable.

R_1: the line resistance.
R_2: the circuit protective conductor resistance.
r_1: the recorded end to end resistance of the line loop of a ring final circuit.

r_n: the recorded end to end resistance of the neutral loop of a ring final circuit.

r_2:   the recorded end to end resistance of the cpc loop of a ring final circuit $Zs = Ze + (R_1 + R_2) \times L \times 1.2$ where the ambient temperature is greater than 20°C.

Index